Jürgen Popp, Thomas Mayerhöfer (Eds.)

Micro-Raman Spectroscopy

Also of interest

Vibrational Spectroscopy
Guozhen Wu, 2019
ISBN 978-3-11-062223-2, e-ISBN 978-3-11-062509-7

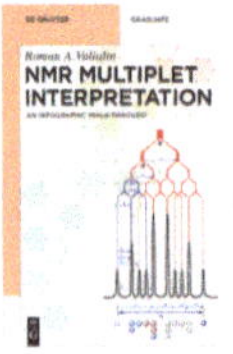

NMR Multiplet Interpretation – An Infographic Walk-Through
Roman Valiulin, 2019
ISBN 978-3-11-060835-9, e-ISBN 978-3-11-060840-3

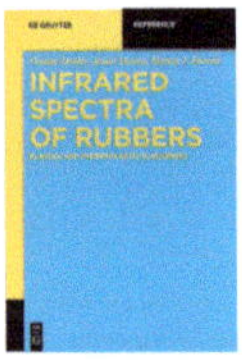

Infrared Spectra of Rubbers, Plastics and Thermoplastic Elastomers
Davies, Davies, Forrest, 2020
ISBN 978-3-11-064408-1, e-ISBN 978-3-11-064575-0

Elemental Analysis – An Introduction to Modern Spectrometric Techniques
Schlemmer, Balcaen, Todolí, Todolí, 2019
ISBN 978-3-11-050107-0, e-ISBN 978-3-11-050108-7

Micro-Raman Spectroscopy

Theory and Application

Edited by
Jürgen Popp and Thomas Mayerhöfer

DE GRUYTER

Editors
Prof. Dr. Jürgen Popp
Friedrich-Schiller-Universität Jena
Institut f. Physikalische Chemie
Helmholtzweg 4
07743 Jena
Leibniz-Institut für Photonische
Technologien e.V.
Albert-Einstein-Str. 9
07745 Jena
Juergen.Popp@leibniz-ipht.de

Priv.-Doz. Dr. Thomas Mayerhöfer
Leibniz-Institut für Photonische
Technologien e.V.
Albert-Einstein-Str. 9
07745 Jena
Thomas.Mayerhoefer@leibniz-ipht.de

ISBN 978-3-11-051479-7
e-ISBN (PDF) 978-3-11-051531-2
e-ISBN (EPUB) 978-3-11-051531-2

Library of Congress Control Number: 2019955231

Bibliographic information published by the Deutsche Nationalbibliothek
The Deutsche Nationalbibliothek lists this publication in the Deutsche Nationalbibliografie; detailed bibliographic data are available on the Internet at http://dnb.dnb.de.

Typesetting: Integra Software Services Pvt. Ltd.
Printing and binding: CPI books GmbH, Leck
Cover image: yuriz/iStock/Getty Images Plus

www.degruyter.com

Contents

List of contributing authors

Yasmine Ahmed
University at Albany, SUNY
Chemistry
Albany, NY
United States
yahmed@albany.edu

Philippe de Bettignies
HORIBA Scientific
R&D
Villeneuve d'Ascq
philippe.debettignies@horiba.com

Thomas Bocklitz
Friedrich-Schiller-University
Institute of Physical Chemistry and Abbe
Center of Photonics (IPC)
Jena
Germany
Thomas.Bocklitz@uni-jena.de

Dana Cialla-May
Leibniz-Institut fur Photonische Technologien
Spectroscopy/Imaging
Jena
Germany
dana.cialla-may@leibniz-ipht.de

Marisia A. Fikiet
University at Albany, SUNY
Chemistry
Albany, NY
United States
mfikiet@albany.edu

Shuxia Guo
Friedrich-Schiller-University
Institute of Physical Chemistry and Abbe
Center of Photonics (IPC)
Jena
Germany
Shuxia.Guo@uni-jena.de

Lenka Halámková
University at Albany, SUNY
Chemistry
Albany, NY
United States
lhalamkova@albany.edu

Izabella-Jolan Jahn
Leibniz-Institut fur Photonische Technologien
Spectroscopy and Imaging
Jena
Germany
izabella.jahn@leibniz-ipht.de

Nathalie Jung
Goethe University
Institute of Pharmceutical Technology and
Buchmann Institute for Moelcular Life Sciences
Frankfurt am Main
Germany
n.jung@em.uni-frankfurt.de

Shelby R. Khandasammy
University at Albany, SUNY
Chemistry
Albany, NY
New York
skhandasammy@albany.edu

Christoph Krafft
Leibniz-Institut fur Photonische Technologien
Spectroscopy and Imaging
Jena
Germany
christoph.krafft@leibniz-ipht.de

Markus Lankers
MIBIC GmbH & Co KG
R&D
Berlin
Germany
markus.lankers@mibi-c.com

https://doi.org/10.1515/9783110515312-203

Lydia Lehniger
Leibniz-Institut fur Photonische Technologien
Spectroscopy and Imaging
Jena
Germany
lydia.lehniger@leibniz-ipht.de

Ewelina M Mistek
University at Albany, SUNY
Chemistry
Albany, NY
United States
emistek@albany.edu

Luc Moens
Ghent University
Department of Chemistry
Ghent
Belgium
luc.moens@ugent.be

Jürgen Popp
Leibniz-Institut fur Photonische Technologien
Spectroscopy & Imaging
Jena
Germany
Juergen.Popp@ipht-jena.de

Anastasia Rousaki
Ghent University
Department of Chemistry
Ghent
Belgium
anastasia.rousaki@ugent.be

Oleg Ryabchykov
Friedrich-Schiller-University
Institute of Physical Chemistry and Abbe Center of Photonics (IPC)
Jena
Germany
Oleg.Ryabchykov@uni-jena.de

Michael Schmitt
Friedrich-Schiller-Universität
Institut für Physikalische Chemie
Jena
Germany
m.schmitt@uni-jena.de

Peter Vandenabeele
Ghent University
Department of archaeology
Ghent
Belgium
Peter.Vandenabeele@UGent.be

Karina Weber
Leibniz-Institut fur Photonische Technologien
Spectroscopy and Imaging
Jena
Germany
karina.weber@leibniz-ipht.de

Maike Windbergs
Goethe University
Institute of Pharmaceutical Technology and Buchmann Institute for Molecular Life Sciences
Frankfurt am Main
Germany
windbergs@em.uni-frankfurt.de

Dana Cialla-May, Michael Schmitt and Jürgen Popp

1 Theoretical principles of Raman spectroscopy

Abstract: This contribution reports on the theoretical foundations of Raman spectroscopy. Since the discovery of the Raman effect in 1928, Raman spectroscopy with its linear and nonlinear variants has established itself as a powerful analytical tool in almost all scientific fields (chemistry, physics, material sciences, pharmacy, biology, (bio)medicine, geology, mineralogy, environmental sciences, etc.). First, a short introduction to linear Raman spectroscopy is given, followed by two approaches to increase the intrinsically weak Raman signal, namely resonance Raman and surface enhanced Raman spectroscopy. The last part of this contribution briefly introduces nonlinear Raman processes observed using pulsed lasers as excitation sources.

Keywords: Raman scattering, resonance Raman spectroscopy, surface enhanced Raman spectroscopy, non-linear Raman spectroscopy

The Raman effect was discovered in 1928 by the Indian physicist C.V. Raman. This method describes the inelastic scattering of photons on a quantized molecular system [1, 2]. In most cases, the vibrational states of molecules are utilized as scattering system and that is why Raman spectroscopy is often referred to as a vibrational spectroscopic technique. Vibrational Raman spectroscopy[1] is a complementary method to IR absorption spectroscopy. The two approaches differ in their physical origin: while IR absorption describes the direct absorption of an IR photon to excite a vibrational quantum (i.e. one photon absorption process), in Raman spectroscopy as mentioned above the vibrational excitation takes place via a two-photon scattering process. In IR absorption spectroscopy vibrational modes leading to a change in the dipole moment during the vibration can be excited, while for Raman active vibrations the polarizability has to change. Since molecular vibrations are distinct for every molecule, vibrational spectra can therefore be interpreted as a type of characteristic "molecular fingerprint" of an examined inorganic, organic or biological molecule or more complex systems like e.g. biological cells and tissue. For the latter vibrational spectral contributions are assigned to proteins, lipids, nucleic acids and

1 From now on, Raman spectroscopy is used synonymously as vibrational Raman spectroscopy.

This article has previously been published in the journal *Physical Sciences Reviews*. Please cite as: Cialla-May, D., Schmitt, M., Popp, J. Theoretical principles of Raman spectroscopy *Physical Sciences Reviews* [Online] **2019**, 4. DOI: 10.1515/psr-2017-0040.

https://doi.org/10.1515/9783110515312-001

carbohydrates. Thus, vibrational spectroscopy is applied for the qualitative and quantitative analysis in chemistry, biology, material and life sciences and biomedicine. However, since water exhibits a large dipole moment its vibrations show a large IR absorption cross-section strongly limiting in vivo studies. This is in stark contrast to Raman spectroscopy where the water vibrations show low scattering cross-sections allowing to record Raman spectra in aqueous solution very easily, thus making Raman spectroscopy a perfect candidate for labelfree *in vivo* investigations on a molecular level [3]. As pathologic anomalies are accompanied by changes in the biochemical composition and structure of biomolecules, the Raman spectrum provides a sensitive and specific fingerprint of the type and state of the specimen. During the last years, Raman spectroscopy has therefore been recognized as an extremely powerful tool for bioanalytical and biomedical applications [4]. Advantages of the technique for biomedical problems include the following: (i) it is label-free and (ii) non-destructive. However, while the molecular selectivity of Raman spectroscopy is very high its sensitivity is very low, i.e. the Raman scattering process is characterized by small Raman cross-sections. In general, only one photon out of 10^8 photons is scattered inelastically. Several Raman signal-enhancing techniques increasing the intrinsically weak Raman scattering cross-sections by several orders of magnitude are known. The two most prominent approaches are resonance Raman spectroscopy and surface enhanced Raman spectroscopy (SERS). In addition to these two linear approaches, non-linear coherent Raman scattering methods, e.g. coherent anti-Stokes Raman scattering (CARS), are also known to providing signal enhancement due to the coherent excitation of vibrational modes. In this paper, a brief theoretical introduction into the most prominent linear and nonlinear Raman techniques will be provided.

1.1 Inelastic light scattering – a brief introduction into linear Raman spectroscopy

The underlying theory of inelastic Raman scattering is comprehensively discussed within the literature [5] to which the interested reader is referred to. The aim of this paper is to give a brief introduction into this topic [6].

1.1.1 Basic principles of Raman spectroscopy

Due to the interaction of light with matter, an electric dipole within molecules is induced since the atoms become polarized, i.e. the center of negative charge is displaced from the atomic nuclei upon the interaction with an electromagnetic field. The induced electric dipole moment is directly proportional to the electric field strength where the polarizability α acts as a proportionality constant. The induced dipole oscillates with the frequency of the external electromagnetic field and acts as a radiation source for an electromagnetic field scattered in all directions. In the case of elastic scattering, i.e. Rayleigh scattering the polarizability α can be seen as static, which is not changing with time, and therefore the frequency of the scattered electromagnetic wave is identical to the frequency of the external electromagnetic field. However, for vibrating molecular systems the polarizability α changes in the course of the molecular vibrations, i.e. the electron density is rapidly adjusted according to the nuclear motions. Therefore, the time-dependent induced dipole moment μ results in:

$$\mu = \alpha_0 E_0 \cos \omega_0 t + \frac{1}{2}\left(\frac{\partial \alpha}{\partial q}\right)\bigg|_{q=0} \cdot q_0 E_0 \cos(\omega_0 - \omega_R)t + \frac{1}{2}\left(\frac{\partial \alpha}{\partial q}\right)\bigg|_{q=0} \cdot q_0 E_0 \cos(\omega_R + \omega_0)t \tag{1.1}$$

Here, the dependency of the polarizability α from the nuclear motion q has been approximated by a Taylor expansion series around the nuclear equilibrium position $q = 0$. As can be seen from eq. (1.1), the induced electric dipole moment μ oscillates at three different frequencies. The first term oscillating at the excitation frequency, ω_0, acts as a radiation source for an electromagnetic source with frequency ω_0, which corresponds to Rayleigh scattering. The second and third terms of eq. (1.1) illustrate oscillations at modulated frequencies, i.e. the difference or sum frequency between the external electromagnetic field ω_0 as well as the frequency of the vibrational mode ω_R, respectively. Here, the second term refers to the Stokes scattering leading to photons with a smaller frequency than the incoming light, which are red-shifted in frequency in comparison with the excitation source. The third term refers to blue-shifted inelastic scattering, i.e. anti-Stokes Raman scattering. As a result, the scattered radiation of a molecule is represented by a superposition of these three frequencies, i.e. ω_0 as well as frequencies associated with the excitation $(\omega_0 - \omega_R)$ or deexcitation $(\omega_0 + \omega_R)$ of vibrational modes of the molecule. Thus, by dispersing the scattered secondary radiation due to the induced dipole moment expressed in eq. (1.1), Rayleigh scattering as well as sidebands, due to inelastic scattering representing the vibrational frequencies of a molecule, are found. In 1928, the appearance of these sidebands was first discovered by C.V. Raman. These sidebands can only be observed if the polarizability is changed along the normal coordinate q with a gradient unequal to zero at the equilibrium geometry, i.e. if $\left(\frac{\partial \alpha}{\partial q}\right)\big|_{q=0} \neq 0$. The Stokes–Raman scattering intensity can be estimated classically by expressing the total power emitted by a Hertzian dipole:

$$I_{Stokes} = constant \cdot I_0 \cdot (\omega_0 - \omega_R)^4 \cdot \left[\left(\frac{\partial \alpha}{\partial q}\right)\bigg|_{q=0}\right]^2 \tag{1.2}$$

This classical description of the inelastic Raman scattering process does, however, not explain the experimentally observed different intensities of Stokes vs. anti-Stokes scattering. Furthermore, such a classical description does not allow understanding rotational or electronic Raman scattering. In order to properly describe Stokes as well as anti-Stokes scattering, a quantum mechanical description is needed. Here, Raman scattering is described as an inelastic scattering process of photons from a quantized molecular system via an extremely short-lived virtual state. This virtual level is considered as a collective quantum energy state of the molecule and the photon as entity during an infinitesimally short time period. Figure 1.1 shows schematic energy level diagrams of Rayleigh-, Stokes- and anti-Stokes-scattering. These diagrams show that Rayleigh and inelastic Raman scattering is a two-photon scattering process via a virtual state.

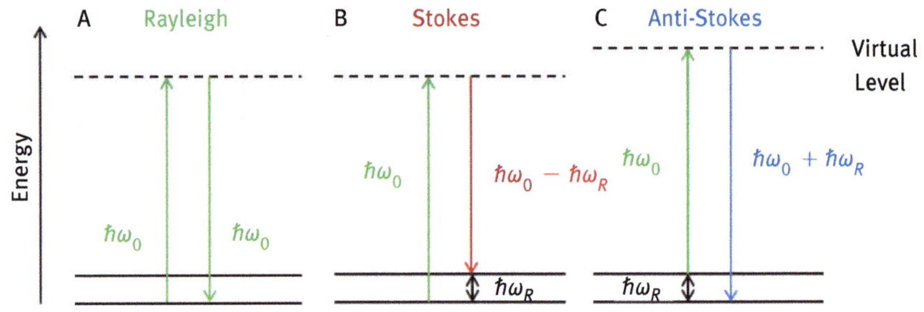

Figure 1.1: Schematic energy diagrams of Rayleigh, Stokes, and anti-Stokes Raman scattering.

Figure 1.1(a) depicts the Rayleigh scattering process where the initial and final scattering states are identical. In the case of Stokes scattering (Figure 1.1(b)) energy is transferred from the photon to the molecule, i.e. the final scattering state is an excited quantum state. In case the molecule originally resides in an excited quantum state, the scattering process can end in the ground state. This process is called anti-Stokes scattering and here energy is transferred from the molecular quantum system to the photon. In the case of vibrational Raman spectroscopy, the depicted quantum states in Figure 1.1 are vibrational levels. According to the Boltzmann distribution, the vibrational ground state is significantly more populated than the energetically excited vibrational states at room temperature. Consequently, the intensity of the Stokes Raman scattering is larger than the intensity of the anti-Stokes Raman scattering. To conclude, the inelastic Raman scattering can be associated with a change in vibrational, rotational or electronic energy, i.e. Raman scattering is performed on quantized systems and the energy is exchanged between the molecules and the incident photons. The exchanged energy is related to the energy levels of vibrational, rotational or electronic transitions. However, as mentioned above the majority of the reported Raman spectroscopic studies are based on the vibrational Raman effect, and thus Raman spectroscopy is often defined as vibrational spectroscopy, being complementary to IR spectroscopy. When plotting Raman spectra, relative wavenumber values are displayed, i.e. the shift of the inelastic scattered light in relation to the applied excitation wavelength (in contrast, IR spectra are plotted in absolute wavenumber values). As already mentioned, the Stokes Raman vibrational modes have a higher intensity than the anti-Stokes Raman lines, and thus the Stokes Raman spectrum is usually recorded. To be able to investigate samples with a spatial resolution in the sub-micrometer range, Raman spectroscopy can easily be combined with conventional microscopy. By means of Raman microspectroscopy molecular maps of heterogeneous samples, e.g. tissue, biological cells or other materials, are generated. As a consequence, Raman imaging allows for recording spectral fingerprint information with a spatial resolution down to the subcellular level when biological material is to be investigated [3]. Raman spectroscopy is known as a spectroscopic method with a very high amount of molecular information; however, as mentioned above its sensitivity is rather poor. Roughly one photon out of 10^8 photons is scattered inelastically, and thus the integration times can become very long as well as the observation of molecules and substances in low concentrations is limited. In order to discuss enhancement strategies of Raman spectroscopy eq. (1.2) needs to be considered. This equation illustrates that the Stokes Raman intensity is dependent on the electronic polarizability $\left(\frac{\partial\alpha}{\partial q}\right)\Big|_{q=0}$ on the light intensity I_0 as well as the excitation frequency ω_0. The two commonly applied methods to enhance Stokes–Raman scattering by several orders of magnitude utilizing the frequency dependency of the polarizability α and increasing I_0 via utilizing plasmonic nanostructures are resonance Raman scattering (RRS) and surface enhanced Raman spectroscopy (SERS). In the case of RRS, the vibrational modes of resonantly excited electronic chromophores are selectively enhanced, whereas SERS is based on the interaction of analyte molecules with nanostructured plasmonic-active metal surfaces and the high electric field intensities generated in their vicinities. In the following section, these Raman signal-enhancing methods are briefly explained.

1.1.2 Resonance Raman scattering (RRS)

RRS is described semi-classically, i.e. the molecule is treated quantum mechanically while the electromagnetic field is considered classically. Therefore, the term $\left[\left(\frac{\partial\alpha}{\partial q}\right)\Big|_{q=0}\right]^2$ from eq. (1.2) is replaced by $|\alpha|^2$, the square modulus of the transition polarizability:

$$I_{Stokes} = constant \cdot I_0 \cdot (\omega_0 - \omega_R)^4 \cdot |\alpha|^2 \tag{1.3}$$

The transition polarizability can be derived by second-order perturbation theory. Following this route, Kramers, Heisenberg and Dirac found an expression for the transition polarizability tensor, which is explained in detail in the literature [6]. A detailed description of the Kramers–Heisenberg–Dirac picture would go beyond the scope of this brief introduction. Only briefly, in case the Raman excitation frequency ω_0 is in resonance with an electronic transition, the transition polarizability can be expressed as follows:

$$(\alpha_{\rho\sigma})_{fi} = \sum_r \left\{ \frac{f|\mu_\rho|rr|\mu_\sigma|i}{\hbar\omega_{ri} - \hbar\omega_0 - i\Gamma_r} \right\}. \tag{1.4}$$

where ρ and σ are the direction *x,y,z* of the incoming and scattered electromagnetic field, respectively. Here the numerator contains transition dipole moments connecting the initial |i⟩ or final |f⟩ state with the manifold of electronic resonant eigenstates of the molecule |r⟩ (see Figure 1.2). From eq. (1.4), it can be seen that the energy denominator becomes small in case the excitation frequency ω_0 matches a molecular electronic transition ω_{ri}. The damping terms Γ avoid that the energy denominator approaches infinity for electronic resonant excitation. This damping term arises due to the finite lifetime of the resonant intermediate states.

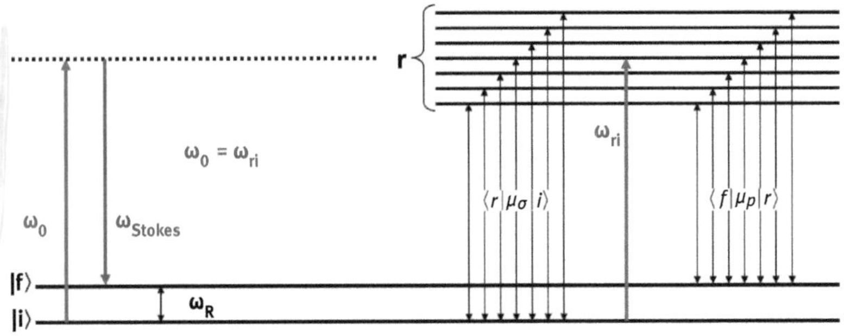

Figure 1.2: Schematic representation of RRS.

Overall, in case the scattering process takes place via a real electronic eigenstate (RRS) instead of a virtual state the transition polarizability tensor is increased dramatically leading to a Raman signal enhancement of up to six orders of magnitude (see also eq. (1.4)). A detailed evaluation of the transition dipole matrix elements in the numerator of eq. (1.4) taking into account the Born-Oppenheimer-, Herzberg-Teller- and Condon-approximation reveals that resonance Raman spectra are dominated by vibrational modes accompanying the resonant electronic transition, i.e. those exhibiting the largest Franck–Condon factors. A detailed quantitative intensity analysis of resonance Raman spectra via eq. (1.4) would allow for the characterization of the excited state geometry within the Franck–Condon region. However, the significance of RRS in biophotonics is rather due to the increased Raman scattering intensity in comparison with non-resonant Raman spectroscopy. Resonance Raman spectra are less complex than non-resonant spectra, which is attributed to the fact that only Raman modes are resonantly enhanced within the RRS spectrum, which are associated with the electronic transition, i.e. which are Franck–Condon active. Thus, based on an excitation of selected electronic chromophores within a biomolecule or an entire cell by using appropriate resonant excitation wavelengths, a chromophore selective and therefore site-specific investigation becomes possible, making RRS an excellent tool in biochemistry, biology and medicine to selectively study electronic chromophores, which might play a pivotal role in large biomolecules. However, also drawbacks of the method need to be mentioned. Due to the excitation with laser light in electronic resonance, photodegradation processes occur within the sample, and thus the sample needs to be moved very fast under the illuminating beam in order to avoid degradation effects. As a consequence, the combination of RRS with microscopy to perform imaging of biological samples is challenging. The largest drawback, however, associated with RRS is the simultaneous excitation of fluorescence. Due to the larger cross-section for fluorescence the Raman scattered light is often overlapped by a broad and strong fluorescence signal. Especially for small Stokes shifts, i.e. the energy difference between the absorption and emission maximum, this issue occurs. In the case of large Stokes shift, the resonance Raman scattered light and the fluorescence signal are well separated. As a strategy to spectrally separate the RRS fingerprint information from fluorescence, excitation wavelengths in the deep UV region (<250 nm) to excite high-lying electronic states can be applied. Since fluorescence originates from the vibrational ground state of the first electronic excited state and RRS occurs from the Franck–Condon point of the resonant high-lying electronic state, the fluorescence signal and the RRS signals are energetically well separated. Finally, the different time scales of RRS and fluorescence could be employed for separation, which is within a time window of 5–10 fs in the case of RRS as well as on the nanosecond scale for fluorescence processes after the electronic excitation. Thus, ultrafast Kerr shutters can be applied allowing for the observation of RRS spectra without the fluorescence background.

1.1.3 Surface enhanced Raman spectroscopy (SERS)

As a further strategy to enhance the Raman scattered light, SERS became very attractive in bioanalytics due to its high sensitivity and specificity allowing for enhancement factors up to 6–10 orders of magnitude for molecules located in close vicinity to metallic nanostructures [7, 8]. The interested reader is referred to the literature to obtain more detailed information than the here presented brief summary of the SERS effect [6, 9]. As contributions to the overall SERS enhancement, two mechanisms are identified: the electromagnetic and chemical enhancement mechanism. The main contribution to the SERS signal is associated with the electromagnetic enhancement which will be discussed in the following section. Due to the interaction with light, surface plasmon polariton modes are excited on the surface of metallic nanostructures being much smaller than the wavelength of the incoming light. As a result of the plasmon polariton excitation, a strong electromagnetic field is induced on the surface of the nanostructure, showing an evanescent character. Thus, the intensity of Raman modes close to the metallic surface is enhanced by several orders of magnitude since the Stokes–Raman intensity is proportional to I_0 (see eq. (1.2)), which is equal to the square of the incident electromagnetic field. The Raman scattered light of the Raman modes in close vicinity to the metallic surface is emitted according to the characteristics of a Hertzian dipole. Additionally, the Raman scattered light experiences a further enhancement due to the overlap of the inelastically scattered photon and the plasmon resonance, which is described as emission enhancement or second part of the electromagnetic mechanism. The overall electromagnetic enhancement factor G is defined as follows:

$$G = M_{Loc}(\omega_0) \cdot M_{Loc}(\omega_R) = \frac{|E_{Loc}(\omega_0)|^2}{|E_0(\omega_0)|^2} \cdot \frac{|E_{Loc}(\omega_R)|^2}{|E_0(\omega_R)|^2} \approx \frac{|E_{Loc}(\omega_0)|^2 \cdot |E_{Loc}(\omega_R)|^2}{|E_0|^4} \tag{1.5}$$

where M_{Loc} is the local field enhancement at the frequency of the laser excitation ω_0 as well as at the frequency of the Raman mode ω_R, E_{Loc} is the locally enhanced electromagnetic field and E_0 is the electromagnetic field of the incoming laser light. In case $\omega_0 \approx \omega_R$, a fourth power dependency of the SERS enhancement occurs. Enhancement factors reported for the electromagnetic mechanism are mostly in the range between 4 and 8 orders of magnitude. As a second contribution to the overall SERS signal, the chemical mechanism is described as the sum of various phenomena: ground-state enhancement due to the interaction of a molecule and the metallic nanoparticle, signal enhancement due to the excitation of a charge transfer processes between a molecule and the metallic surface, and the resonance Raman enhancement. The enhancement factor of the chemical enhancement is within the range of 10^1 to 10^3. It is observed that the SERS spectrum can differ from the Raman spectrum of the same substance. Thus, not all Raman modes are enhanced by the same factor. This is attributed to the surface selection rules, which shows that the modes perpendicular to the metallic surface are enhanced whereas the modes parallel to the surface are not enhanced or of very weak intensity. Moreover, the evanescent character of the electromagnetic field on the surface of the metallic nanostructures is responsible for the breakdown of the conventional spectroscopic selection rules, and thus forbidden Raman modes might be observable under SERS conditions (field gradient Raman effect).

All the aforementioned Raman approaches can be considered as linear, where the Raman scattering intensity scales linear with the excitation intensity. However, various non-linear Raman scattering processes exist, which will be briefly explained in the following section.

1.2 Nonlinear Raman-based processes

In this section, nonlinear Raman processes are highlighted and briefly discussed. A detailed description of nonlinear Raman techniques can be found in the literature [6, 10].

1.2.1 Hyper-Raman spectroscopy

Due to the application of intense light sources, a nonlinear polarization occurs. As a consequence, the dipole moment μ is no longer proportional to the electric field E applied to the system. The nonlinearity can be illustrated as follows:

$$\mu = \alpha \cdot E + \frac{1}{2}\beta \cdot E^2 + \frac{1}{6}\gamma \cdot E^3 + \ldots \tag{1.6}$$

where β is understood as the first hyperpolarizability for which $\beta \ll \alpha$ is valid and γ with $\gamma \ll \beta$ being the second hyperpolarizability. As mentioned earlier, the polarizability α depends on the nuclear geometry, i.e. on motions of the nuclei, within the molecule. This is also true for the first, second and higher hyperpolarizability terms. It is found that the second-order nonlinear-induced dipole moment $\mu^{(2)} = \frac{1}{2}\beta \cdot E^2$ can oscillate at four different frequencies: $2\omega_0$ which is described as the hyper-Rayleigh scattering; $(2\omega_0 + \omega_R)$ and $(2\omega_0 - \omega_R)$ as so-called anti-Stokes hyper-Raman and Stokes hyper-Raman scattering, respectively; and at ω_R as a vibrational mode within the IR spectral range. In other words, within the hyper-Raman process a two-photon transition into the virtual states takes place and for those molecules relaxing back into the initial molecular state, hyper-Rayleigh scattering is observed. In the case of a relaxation back into an energetically excited level, Stokes hyper-Raman scattering occurs. The anti-Stokes hyper-Raman scattering is achieved when the two-photon transition into the virtual level starts from an energetically excited state and the molecule relaxes back to the vibrational ground level. According to the Boltzmann distribution, the anti-Stokes hyper-Raman scattering is less intense than its Stokes counterpart. It is important to mention that these hyper Raman scattering frequency components can only be observed when applying intense laser sources e.g. ultrashort laser pulses. Thus, hyper-Raman scattering plays only a minor role in Biophotonics or also other scientific disciplines. The combination of hyper-Raman spectroscopy with the high electromagnetic field intensity in close vicinity to nanostructured metallic surfaces (creating surface enhanced hyper-Raman spectroscopy, SEHRS) is resulting in complementary spectroscopic information due to different selection rules and a very strong enhancement based on plasmonic and nonlinearity effects [11].

1.2.2 Coherent anti-stokes Raman spectroscopy (CARS) and CARS microscopy

All Raman-based scattering processes introduced so far are considered as spontaneous processes, i.e. the vibrations of all molecules within the laser focus are excited independently (i.e. oscillate with random phase relationships) for employing a single frequency of the incoming light ω_0. Thus, a spontaneous Raman spectrum is a superposition of all excited vibrations of the molecules based on the application of a single incoming frequency. In this section, coherent anti-Stokes Raman scattering (CARS) as the most prominent example of a coherent Raman process is introduced. Here, the molecules are excited by two laser fields having different frequencies, i.e. the laser frequency ω_L and the Stokes frequency ω_S. In case the frequency difference between ω_L and ω_S matches a molecular vibrational frequency ω_R, all molecules within the common focus of the two laser fields oscillate coherently. To be more precise, in CARS three laser pulses are involved to generate a coherent and spatially directed CARS signal, whereas two laser pulses with the laser frequency ω_L and the Stokes frequency ω_S are interacting with the sample of interest. In case the frequency difference between ω_L and ω_S corresponds to a Raman transition ω_R (i.e. $\omega_L - \omega_S = \omega_R$) the molecules are excited to coherent in-phase oscillations. Subsequently, another photon with the frequency ω_L is scattered inelastically off the ensemble of coherently excited modes by emitting an anti-Stokes signal with the frequency ω_{aS}:

$$2\hbar\omega_L - \hbar\omega_S = \hbar\omega_{aS} \tag{1.7}$$

In other words, whenever the difference frequency equals that of a Raman transition, a CARS signal can be recorded. In total, CARS can be considered as four-photon process (see Figure 1.3(a)).

The frequency of the anti-Stokes Raman signal results from the energy conservation (see eq. (1.7)), while the direction of the coherent CARS signal is determined by the wave vector conservation. The sum of the wave vectors of the four photons involved in the process must be zero for the CARS signal to become maximum:

$$\Delta k = k_{aS} - (2k_L - k_s) = 0 \tag{1.8}$$

Thus, this phase matching condition is achieved by special arrangements, which is for the investigations of gases a collinear beam arrangement. In condensed phases, a collinear arrangement between four waves is not possible and therefore specially designed non-collinear beam geometries like e.g. a BOXCARS or folded BOXCARS arrangement is applied. Due to its coherent nature, the CARS signal is orders of magnitude more sensitive than the weak spontaneous Raman scattering. By comparing

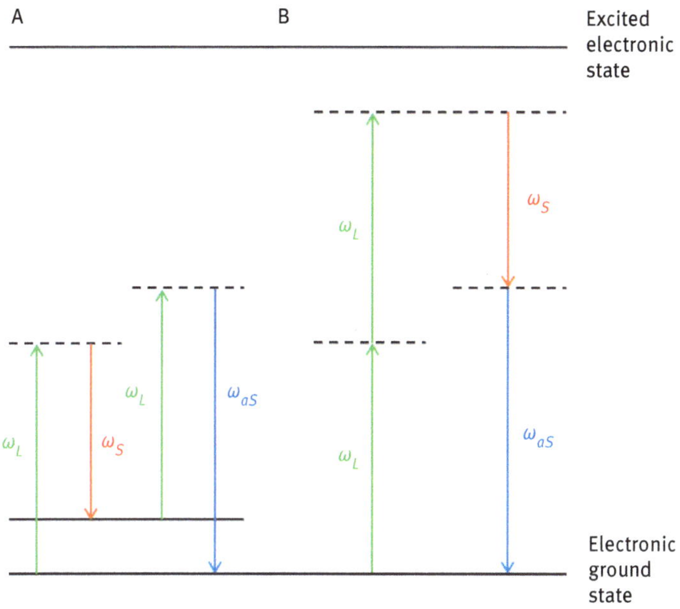

Figure 1.3: Energy level diagram of (A) Raman resonant CARS and (B) non-resonant CARS background.

spontaneous linear Raman spectroscopy with the nonlinear CARS process the following observations can be made: (1) The Raman scattering process is non-directional and incoherent, whereas CARS is associated with a directed and coherent signal. Thus, the Raman scattering intensity is significantly enhanced. (2) The Stokes–Raman information is red-shifted related to the excitation wavelength and as a consequence, an overlap with a disturbing fluorescence signal often occurs. In CARS, the signal is blue-shifted and thus no fluorescence background is disturbing the measurements. (3) The Raman scattering intensity scales linearly with the number of molecules N, while for CARS a quadratic dependency N^2 is obtained. (4) In linear Raman spectroscopy, an energy exchange between the scattered photon and the molecule occurs. The CARS process can be interpreted as a series of Stokes and anti-Stokes processes where the molecule ends up in the initial molecular state (see Figure 1.3(a)). (5) The line shape of Raman modes is easy to interpret, i.e. mostly Voigt profiles are present. In contrast, the situation under CARS conditions is very complex. When recording CARS spectra by tuning ω_S for fixed ω_L and recording the CARS intensity often complex CARS spectral lineshapes are observed, not corresponding to the observed linear Raman spectra. Beside the Raman resonant CARS contribution, which is based on Raman transitions i.e. molecular vibrations (see Figure 1.3(a)), one must also consider a non-resonant CARS part, which is usually caused by off-resonant transitions due to the electronic structure of the medium (see

Figure 1.3(b)). This non-resonant CARS background is real and does not show a strong frequency dependence (i.e. can be regarded as constant in the proximity of the vibrational resonance ω_R). The interference between these two contributions, i.e. Raman resonant and non-resonant part, often leads to asymmetric CARS line shapes, which are difficult to interpret. The situation becomes even more complicated when Raman transitions are very close or spectrally overlapping. Only for isolated Raman transitions with large scattering cross-sections, the observed CARS lineshape is similar to that observed in linear Raman spectroscopy. Nonetheless, since CARS combines a Raman signal-enhancing process due to the coherent excitation of molecular vibrations with directed emission, and no disturbing fluorescence background, this nonlinear variant of Raman spectroscopy became very attractive in Biophotonics as microscopic contrast phenomenon. When implementing a CARS microscope, two collinear laser beams (L and S) are focused on a point in the sample via a microscope objective with the largest possible aperture angle. When using a microscope objective, i.e. strongly focused laser pulses, phase matching is uncritical, since the focus for each wave vector component of the laser beam L contains the Stokes beam S required for phase matching and the CARS signal is only generated over a very short interaction length. CARS microscopy allows for real-time imaging of characteristic Raman peaks with large Raman scattering cross-sections, i.e. polar molecular groups (e.g. CH_2 stretch vibrations) over large areas up to cm^2 at diffraction limited spatial resolution. Finally, the excitation source for CARS microscopy needs to be chosen with care. Large field strengths are required due to the nonlinearity of the process. Thus, ultrashort laser pulses are required. The employed pulse length should be as short as possible for an efficient CARS signal; however, long enough that the spectral width of the laser pulse approximately matches the line width of a Raman transitions. It was shown that for CARS microscopy, picosecond pulses are the best compromise [12]. As mentioned above, CARS is not background-free and that is why CARS images recorded for Raman resonances with low scattering cross-sections will contain a non-resonant background significantly reducing the molecular contrast. Furthermore, since the CARS intensity scales quadratically with the number of molecules a quantitative analysis of CARS images becomes often challenging. These drawbacks can be circumvented by using another coherent Raman method, namely stimulated Raman scattering (SRS) microscopy, which is however experimentally more complex compared to CARS. SRS is background-free and exhibits a linear dependency on the concentration. In the following section, the basic concept of SRS will be briefly summarized.

1.2.3 Stimulated Raman scattering (SRS)

As a second coherent Raman process, SRS is considered which is based on an optical amplification of the longer wavelength beam ω_S when two laser beams with a frequency of ω_L and ω_S are guided through a Raman-active sample and their frequency difference matches a Raman transition, i.e. $\omega_L - \omega_S = \omega_R$. To be specific, the incident laser beam loses an energy quantum with the energy of $\hbar\omega_L$ whereas the Stokes beam gains an energy quantum $\hbar\omega_S$. As a consequence, two Stokes photons with the frequency ω_S are generated (see Figure 1.4).

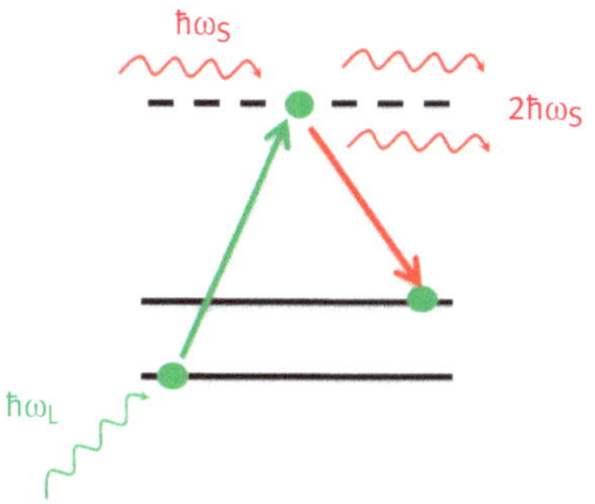

Figure 1.4: Energy level diagram of stimulated Raman scattering.

In SRS, the investigated molecules interact simultaneously with two light waves with the frequencies $\omega_L - \omega_S = \omega_R$. Thus, the waves are coupled by the molecule vibrating at the frequency $\omega_R = \omega_L - \omega_S$ and an energy exchange between the laser wave and the Stokes wave occurs resulting in a stimulated Raman loss ΔI_L in the laser beam intensity I_L and a stimulated Raman gain ΔI_S in the Stokes beam intensity I_S:

$$\Delta I_S \propto N \cdot \left(\frac{\partial \alpha}{\partial q}\right)_0 \cdot I_L \cdot I_S \tag{1.9}$$

$$\Delta I_L \propto -N \cdot \left(\frac{\partial \alpha}{\partial q}\right)_0 \cdot I_L \cdot I_S \tag{1.10}$$

The intensity of the SRS signal (i.e. ΔI_L or ΔI_S) scale linear with concentration (N) and are background free since a loss or a gain in the Laser or Stokes beam intensity is only generated if $\omega_L - \omega_S$ matched a Raman resonance ω_R. However, SRS is experimentally more challenging since the real time detection of a small signal (i.e. laser loss or Stokes gain) on top of a large laser background requires sensitive lock-in detection schemes and fast tunable laser sources.

References

[1] Raman CV, Krishnan KS. A new type of secondary radiation. Nature (London, United Kingdom). 1928;121:501–2.

[2] Landsberg G, Mandelstam L. A novel effect of light scattering in crystals. Naturwiss. 1928;16:557–8.

[3] Krafft C, Schmitt M, Schie IW, Cialla-May D, Matthäus C, Bocklitz T, et al. Label-free molecular imaging of biological cells and tissues by linear and nonlinear Raman spectroscopic approaches. Angew Chem Int Ed. 2017;56:4392–430.
[4] Krafft C, Schie IW, Meyer T, Schmitt M, Popp J. Developments in spontaneous and coherent Raman scattering microscopic imaging for biomedical applications. Chem Soc Rev. 2016;45:1819–49.
[5] Schrader B. Infrared and Raman spectroscopy: methods and applications. Weinheim: Wiley-VCH Verlag GmbH, 1995.
[6] Popp J, Tuchin VV, Chiou A, Heinemann SH. Handbook of biophotonics. vol.1: basics and Techniques. Weinheim, Germany: Wiley-VCH Verlag GmbH & Co. KGaA, 2011.
[7] Cialla-May D, Zheng X-S, Weber K, Popp J. Recent progress in surface-enhanced Raman spectroscopy for biological and biomedical applications: from cells to clinics. Chem Soc Rev. 2017;46:3945–61.
[8] Jahn IT, Žukovskaja O, Zheng X-S, Weber K, Bocklitz TW, Cialla-May D, et al. Surface-enhanced Raman spectroscopy and microfluidic platforms: challenges, solutions and potential applications. Anal. 2017;142:1022–47.
[9] Jahn M, Patze S, Hidi IJ, Knipper R, Radu AI, Mühlig A, et al. Plasmonic nanostructures for surface enhanced spectroscopic methods. Anal. 2016;141:756–93.
[10] Boyd RW. Nonlinear optics. New York: Academic Press, 1992.
[11] Madzharova F, Heiner Z, Kneipp J. Surface enhanced hyper Raman scattering (SEHRS) and its applications. Chem Soc Rev. 2017;46:3980–99.
[12] Gottschall T, Meyer T, Schmitt M, Popp J, Limpert J, Tünnermann A. Advances in laser concepts for multiplex, coherent Raman scattering micro-spectroscopy and imaging. TrAC Trends Analyt Chem. 2018;102:103–9.

Philippe de Bettignies

2 Optics/instrumentation

Micro-Raman spectroscopy: theory and application

Abstract: Raman spectroscopy has undercome a major development during the last decades, making it more and more popular. One of the reason is that the instrumentation has become more accessible, by making the Raman microscopes much more affordable, reliable and user-friendly. Although the fundamental design has not really changed since the origins, it is important to know what the different components are made of and what is their purpose. Is that sense this chapter about the Optics of the Raman microscope will help to understand how the instrument works and what are the specifications and the useful characteristics.

Keywords: Raman microscopy, Raman spectroscopy, Raman imaging, Laser, Confocal microscope, Spectrometer, Spectral resolution, Spatial resolution

2.1 Introduction

First introduced in the mid-1970s, the Raman micro-spectrometer or micro-Raman system or Raman microscope (see Figure 2.1) opened up a whole new dimension of spectroscopic analysis. Since that time the Raman micro-spectrometers have significantly evolved. They are now far easier to operate, and laser adjustment and alignment are virtually eradicated. It becomes a simple operation to use the micro Raman instrument with computer-controlled operations like laser switching and grating selection.

Although their design has not fundamentally changed since the beginning, there is a number of details that must be known, in order to understand how they work. The goal of this chapter is to make clear all main technical aspects of the components of a Raman Micro-Spectrometer, to be able to understand a technical specification, a commercial brochure, a quote or an experiment using such instrument.

2.2 Overall scheme

The Micro Raman system follows the usual rule of any scientific instrumentation: there is an excitation source, a sampling unit, and a detection part. In our case the

This article has previously been published in the journal *Physical Sciences Reviews*. Please cite as: de Bettignies, P. Optics/instrumentation *Physical Sciences Reviews* [Online] **2019**, 4. DOI: 10.1515/psr-2018-0027

https://doi.org/10.1515/9783110515312-002

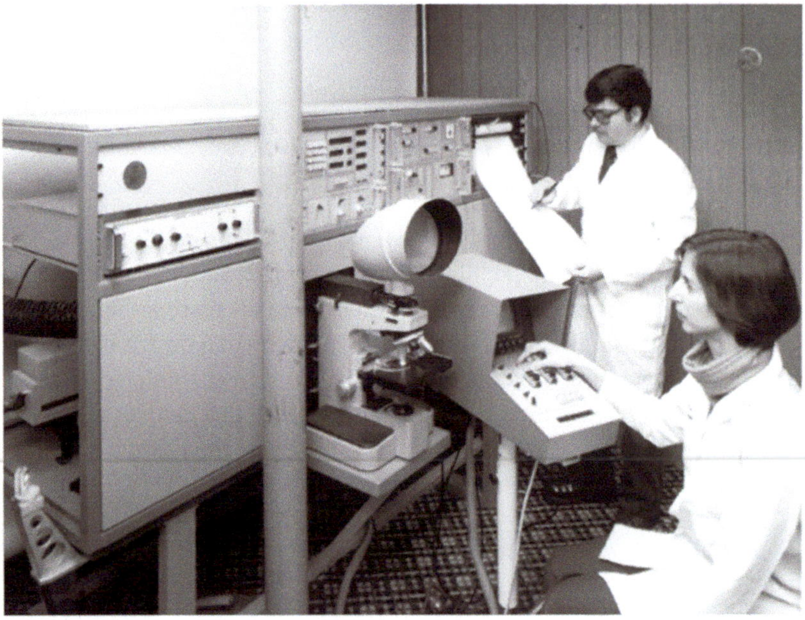

Figure 2.1: "Mole" first micro-Raman system in the late 70's.

excitation source is the monochromatic light, provided by a laser. The sampling unit is a microscope, very close to a conventional light microscope. And the detection part is a spectrometer.

A Raman microscope combines a Raman spectrometer with a standard optical microscope. The excitation laser beam is focused through the microscope to create a tiny spot with a diameter in the order of a micrometer. The Raman signal from the sample is collected from a similar area, passes back through the microscope into the spectrometer and is there analyzed for spectral information.

All Micro Raman systems today have roughly the same arrangement, and can easily be described by following the path of the light in the instrument in Figure 2.2:

1. One or several laser and coupling optics with accessories to control their properties.
2. A microscope, with video capabilities, one or several objectives and a sample holding/scanning unit.
3. A filtering unit, to isolate the Raman scattered light from the laser scattered light (remember that the Raman effect is really weak, at least 1 million times weaker than the incoming laser light).
4. A spectrometer, to disperse the light using a diffraction grating and to reveal the Raman spectrum.

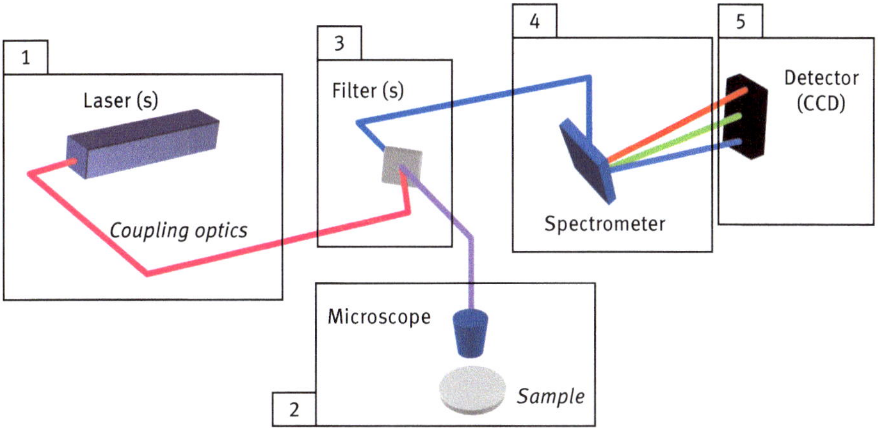

Figure 2.2: Micro-Raman system overview.

5. A detector integrated into the spectrometer, to digitize the spectrum. It's usually a CCD camera.
6. And don't forget a computer, to control the instrument and display the spectra, images and other analysis results.

2.3 Laser

Lasers are a very important part of the Raman system. Their performance will directly influence the quality of your result, especially in terms of spectrum quality and spatial resolution. That's why the Raman manufacturers carefully select them and usually provide them.

2.3.1 Wavelength choice (and why)

The correct selection of the laser wavelength is an important consideration for Raman spectroscopy. Often several laser wavelengths may be employed to achieve the best detection of the Raman signal versus background noise.

2.3.1.1 Fluorescence

First, it can happen that the sample, especially of an 'organic' or 'biological' nature, will be quite fluorescent species. Exciting these samples with a laser in the green (532 nm) may promote this fluorescence and may swamp any underlying Raman spectrum to such an extent that it is no longer detectable.

In this instance, the use of a laser in the red (633 nm) or NIR (785 nm) may provide a solution, as shown in Figure 2.3. With lower photon energy, a red or NIR laser may not excite the electronic transition (and hence the fluorescence), so the Raman scatter may be far easier to detect. Or, if the emission is in the green-yellow-red region, using a much shorter wavelength like blue (473 nm), violet (405 nm) or even down to UV (325 nm or 266 nm) will allow to detect Raman spectrum before (wavelength-wise) the fluorescence spectrum.

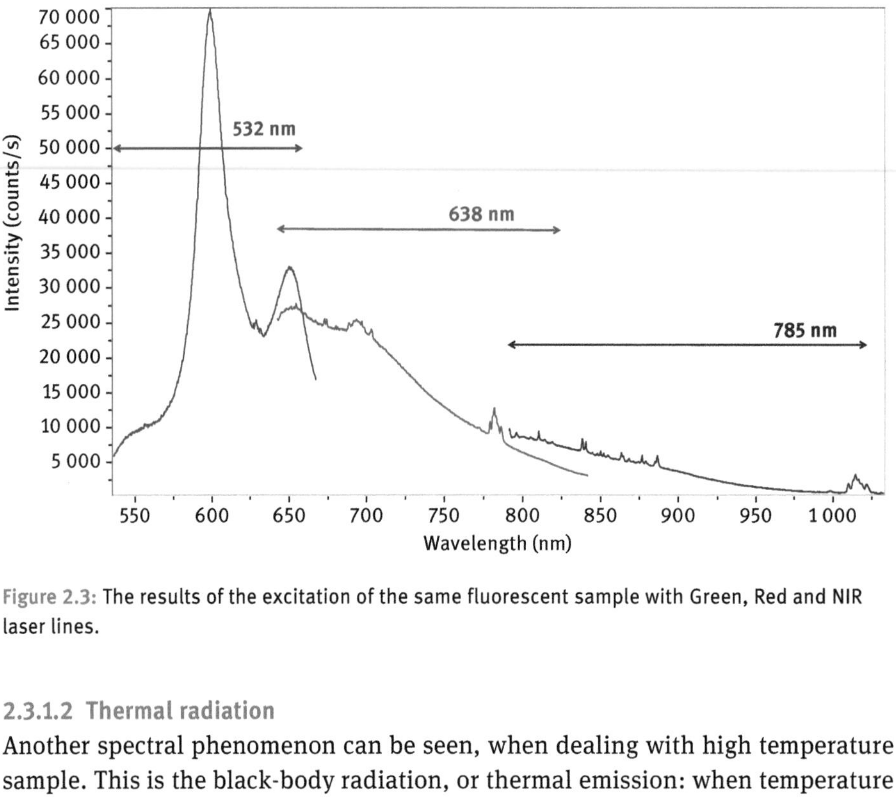

Figure 2.3: The results of the excitation of the same fluorescent sample with Green, Red and NIR laser lines.

2.3.1.2 Thermal radiation

Another spectral phenomenon can be seen, when dealing with high temperature sample. This is the black-body radiation, or thermal emission: when temperature rises, the thermal emission increases and its peak shift towards lower wavelengths, which can completely mask the Raman spectrum, like in the previous case of fluorescence. The theoretical black-body spectral emission is shown in Figure 2.4.

2.3.1.3 Raman scattering efficiency

Increasing the wavelength, from green to red to NIR, will decrease the scattered signal level, because Raman scattering efficiency is proportional to $1/\lambda^4$. An example of such effect is given at Figure 2.5. As a consequence longer integration times or higher power lasers are usually required when working in NIR region.

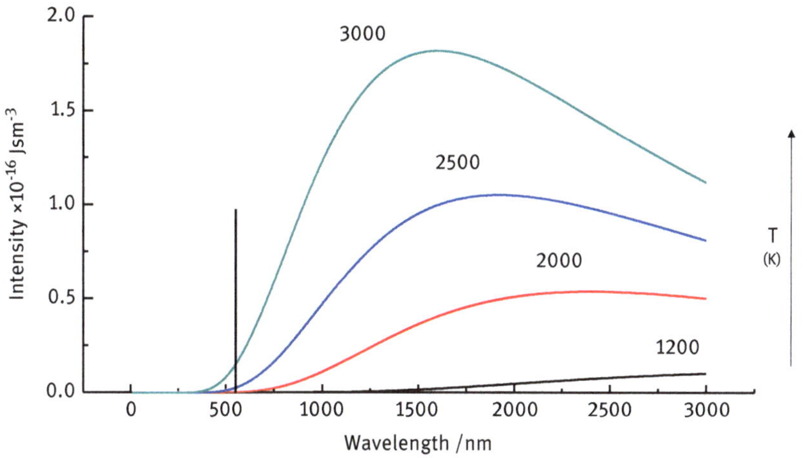

Figure 2.4: Black body radiation spectrum as a function of temperature.

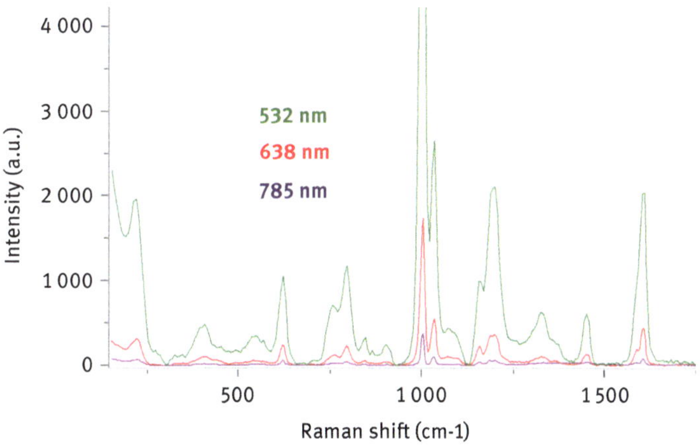

Figure 2.5: Raman efficiency versus excitation wavelength (Polystyrene spectra, normalized for acquisition time, laser power, detector response).

2.3.1.4 Light penetration depth

For some materials like silicon (or other semiconductors, see Table 2.1), the penetration depth will depend on the laser wavelength, allowing to probe below the surface by choosing the appropriate wavelength. This is also true for biological materials like skin or other tissues.

2.3.1.5 Spatial resolution/spot size

As it will be described later in this chapter (see Section 2.4.5), the laser spot size that will directly affect spatial resolution is proportional to the laser wavelength.

Table 2.1: Penetration depth vs. laser wavelength.

Laser wavelength (nm)	Penetration depth in Si (nm)	Penetration depth in Ge (nm)
785	10,000	200
633	3000	32
514	762	19.2
488	569	19
457	313	18.7
325	~10	~9
244	~6	~7

Roughly speaking a 785 nm lase will have a focused spot twice larger than a 405 nm laser!

Thus, it is often most practical to have a number of laser wavelengths available to match the various sample properties one may encounter, be it resonance enhancements, penetration depth or fluorescence.

2.3.2 Laser types

The most popular laser wavelengths are 532 nm (green) and 633 nm (red), for excitation in the visible range, and 785 nm of excitation in the near infrared.

These laser are actually of different types, each of them with different properties and available wavelength (see Table 2.2 and Table 2.3):

- Gas laser: an electric discharge through a gas medium, inside a sealed cavity. They tend to become obsolete due to the arrival of the other advantageous technologies
- Laser Diode: a semiconductor emitter, which facets have been coated to create a laser cavity.
- Diode Laser: adding to a laser diode some external components to have better quality (spectral linewidth and stability, beam profile)
- DPSS: "diode pumped solid-state". A laser diode is used to pump the emitter crystal (most popular is Nd:YAG), located inside the laser cavity. Inside (or sometimes outside) the cavity are placed other components to double (or modify) the light frequency in order to generate the appropriate wavelength. They have intrinsically narrow linewidth, but need to be stabilized
- Fiber laser: the amplification medium is a fiber. Can generate higher power, with a very long lifetime

Speaking of the lifetime, it's usually for a laser diode, a diode Laser or a DPSS around 10,000 h. For a gas laser it's much less, around 2000 h, and for a fiber

Table 2.2: Popular laser for visible and NIR range and today's main technologies.

Wavelength	Gas laser	Solid state laser	Diode laser	Fiber laser
405 nm (violet)			It's the blue-ray diode	
442 nm (blue)	Helium-Cadmium			
457 nm (blue)	Argon	X		
473 nm (blue)		X		
488 nm (blue)	Argon	X		
514 nm (green)	Argon	X		
532 nm (green)		frequency doubled Nd:YAG		
594 nm (orange)		X		
633 nm (red)	Helium Neon		X	
638–640 nm (red)			X	
660 nm (red)		X		
671 nm (red)		X		
691 nm (red)			X	
785 nm (near IR)			It's the DVD diode	X
830 nm (near IR)			X	
980 nm (IR)			X	
1064 nm (IR)		Nd-YAG		

Table 2.3: Popular laser for UV range.

Wavelength	Gas laser	Solid state laser
364 nm	Argon	
355 nm		tripled Nd:YAG
325 nm	Helium–Cadmium	
266 nm		quadruple dNd:YAG
257 nm	doubled Argon	
244 nm	doubled Argon	
229 nm	doubled Argon	

laser it can reach 100,000 h. Note that if you never turn it off, 10,000 h is only one year of continuous operation!

With respect to the size and price, the gas lasers are more bulky and more expensive (except the HeNe at 633 nm). Then the DPSS are more affordable with a more reasonable footprint, typically less than 20 × 10 × 10 cm. And the diode lasers are the most compact and cheapest, reaching down to 5 × 2 × 2 cm.

The laser technology is evolving pretty fast, and the laser manufacturers are always improving their design and reliability. Especially in this respect the next years will probably show the arrival of fiber-based laser (to differentiate from fibered laser), which show a lifetime of 100,000 h. For the moment their drawback is their high price and their high power (more than 1 W).

2.3.3 Spectral quality

Although there is a huge number of available lasers in the world, not all of them are suitable for Raman spectroscopy and even more specifically Raman microscopy.

The important features for a laser to be qualified for Raman are:

- wavelength stability over time → reproducibility of the spectrum
- Linewidth, and its stability over short time (jitter) → spectral resolution.
  This is often a quality of "single-mode" laser or SLM (for Single Longitudinal Mode)
- Spectral purity → access to low frequencies and long integration time.
  In this respect you may look for ASE suppression ratio (Amplified Spontaneous Emission, a background generated by the laser medium), or Side Mode Suppression ratio (a side mode is another laser wavelength that is emitted simultaneously, very close to the main laser line, in the laser diodes).
  Gas laser will exhibit "plasma lines" (light emitted by the plasma), DPSS can show the "pump diode line" (808 nm). They need to have a narrow bandpass filter (also called laser line filter – see Figure 2.13).

2.3.4 Spatial quality: beam size, divergence, ellipticity

2.3.4.1 Beam parameters

Laser beam are characterized by 2 parameters that are linked: size and divergence. The beam size is usually measured at the laser exit, but will increase with the distance because of the natural divergence of the beam. The special fact about a laser beam, being (in theory) a Gaussian beam, is that if it has a small size, it will highly diverge, and consequently, it must be expanded in order to be collimated (lower divergence). Table 2.4, Figure 2.6 and Figure 2.7 explicits the beam parameters, their name and the relations between them.

The key point is that the laser beam must be adapted to the instrument.

For a typical laser used in Raman microscopes like 633 nm, with a beam waist w0 = 0.5 mm, the previous equation would give theta = 0.4 mrad.

It is then quite usual to have a "beam expander" on the laser path to expand w0 value and therefore reduce the divergence of the beam, in order to perfectly match the microscope optics.

Table 2.4: Gaussian beam parameters and fundamental equation.

Beam radius at $1/e^2$ (13.5 %)		**w**	
At the laser aperture	Initial beam radius at $1/e^2$	w0	Beam waist
	Initial beam diameter	D0 = 2w0	Different from the FWHM (Full Width at Half Maximum)
In the far field	Beam size/diameter	D(z) = 2w(z)	$D(z) = 2\theta z$
	Divergence (half beam)	θ	$\theta = \frac{\lambda}{\pi w_0}$

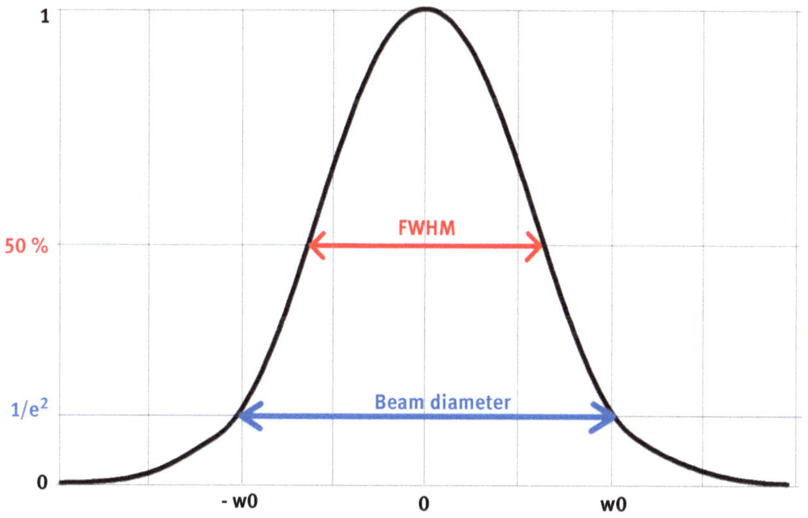

Figure 2.6: Gaussian beam profile of a laser beam.

2.3.4.2 Beam quality

All the previous relationships are only true for a perfect Gaussian beam.It's often the case for microscopy to be able to reach "diffraction limited" performances. But because of the emitting source or some aberrations it can be elliptical or rectangular. In detail, ellipticity will cause:

- Different divergences in X and Y
- An elliptical laser spot when focused
- Different focus spot (astigmatism)

So this should be avoided as much as possible.

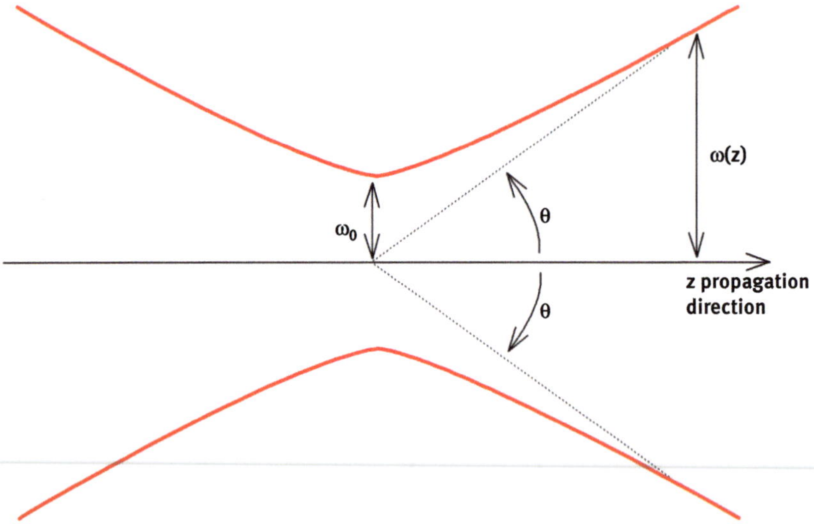

Figure 2.7: Gaussian beam parameters.

More generally speaking, the M^2 factor measures how close to a perfect Gaussian beam the laser is. It should be close to 1 (1.2 is good, 1.5 is acceptable, 2 and above is bad) for microscopy.

Transverse mode describes the shape of the beam. It must be TEM00 for microscopy, it's easy to understand why by looking at the different modes profiles (Figure 2.8).

2.3.5 Power and power control

Usually the laser is specified with their power at the source. But does really matter is the power at the sample. This depends on the coupling efficiency, and on the ability to adjust the power of the laser.

The laser power at sample often needs to be adjusted to protect the sample:

- Not to damage the sample (thermal effects: burn, boil, cook ...)
- Not to modify the properties of the sample (heat will cause changes in stress, crystallinity, conformation ... that can be seen in the spectra as peak shift, broadening ...)

It's also useful when you want to see the sample and the laser simultaneously, as the laser would saturate the camera.

As most laser have fixed power, a device to reduce and adjust the laser power is present in most of the commercial instruments. The most common and easy to use

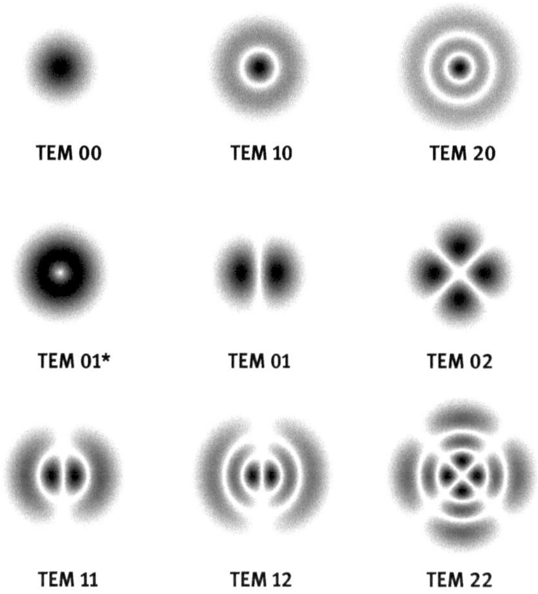

Figure 2.8: Transverse modes profiles.

is a neutral density (ND) filter wheel (or slider), with fixed values of the ND filters (or continuously variable). If the laser is linearly polarized, one can also mention the use of a rotating polarizer (based on Malus law); it must be associated with a half-wave plate in order to keep the polarization state constant.

The attenuation is often expressed in “optical density” following the formula:

$$OD = -\log T \tag{2.1}$$

With T the transmission of the ND filter.

Examples: OD 0.3 = 50 %, OD 1 = 10 %, OD 4 = 0.01 %

Note: in some case the laser power can be adjusted directly within the laser. The drawback of this is generally a lower range of adjustment, and a chance to modify the properties of the laser (emission wavelength, monomode operation, beam position, overall stability).

2.3.6 Laser safety

First safety rule: never try to look directly in the laser beam! Wear protective glasses and avoid direct eye and skin exposure. The optician usually takes his business card (or any white paper) to check that the laser is present or not. You should never get any laser beam in your eye, as this can cause irremediate damage.

In this effect any system using a laser is classified in several safety classes (IEC 60,825 standard) described in Table 2.5:

Table 2.5: Laser class for safety.

Class	Risk	How to use
1	no danger, no dangerous beam can be accessible	no special installation, no glasses
2	no danger unless when deliberate stare into the beam	no special installation, no glasses
3	danger, a dangerous beam is or can be accessible	recommended glasses and special room installation
4	high danger (even for diffused light): high power (>500 mW) and/or invisible (UV or IR)	need glasses + controlled access

In any case, the manufacturer must provide a system which is not dangerous by design, which means that all beams are protected when it's not necessary to have access to it. Usually the only access is on the sample.

Read carefully the user manual of the instrument, as it is mandatory to have in there a section dedicated to laser safety, identifying the laser apertures and giving instructions for a safe use.

Class 1 usually means that the system is fully enclosed, and when the sample access door is opened, a safety switch blocks the laser or switch it off. An indirect benefit of the laser safety enclosure is that it prevents ambient light to perturb the measure.

2.4 Microscope

2.4.1 Types of microscopes

Microscopes are usually classified in three types:

- Regular microscope, with a full frame. The objectives have a fixed position and the focus is made by moving the sample up and down.
- Open microscope, with a detachable or no bottom part. This allows to use larger or heavier sample holders or accessories, like cryostats, heating cooling stages, pressure cells. Usually in this case the focus is made by moving the objectives up and down.
- Inverted microscope, in which the objectives are below the sample. They are mainly used for life science because they give more access for live cell accessories

2.4.2 Illumination techniques and video camera

Whether your sample is transparent or opaque, you will choose different ways of lighting it:

- Reflection illumination or "episcopy" (the light goes through the objective for illumination and is collected via the same path after reflection on the sample). It is used in most cases but is especially meant for diffusing, reflecting or opaque samples.
- Transmission illumination or "diascopy" (the light goes through the condenser for illumination then through the sample and is collected via the objectives) is adapted for transparent samples.

Depending on the optical microscope and the way it is integrated to the Raman instrument, it is possible to add other modalities to the regular "brightfield" (episcopic or diascopic) imaging.

Examples includes, but are not limited to, darkfield, polarized light, DIC, epifluorescence, phase contrast ... Each of this technique being adapted for a specific type of samples by adding other types of contrast.

A video camera (CMOS or CCD) is mandatory in order to perform a measure at the microscopic level, to visualize the sample, see where you are measuring exactly and if you're in focus.

Actually these tasks can also be done by a binocular head, but a camera allows image acquisition with following analysis to show in the instrument software a lot of additional information, like the image scale, the mapping area, distances, profiles ... and of course to record these images and related informations.

The video image can also be used for various options based on image analysis, like autofocus (see Section 2.8 and 2.10.2.1).

There is a tremendous range of video cameras available. The chip size and pixel density influence the frame rate of the video acquisition, which can be observed by moving the sample. In some cases there can be several cameras in the system.

2.4.3 Objectives

The objectives should be chosen based on their use. Raman microscopes are often equipped with a wide field objective (like 5x or 10x) for sample navigation, and an objective to perform the Raman measurement itself, like 50x or 100x objectives.

Some key words that you must know (Figure 2.9):

- *Magnification*: defined by the ratio between the size of the image vs the size of the object, which is also the ratio of the camera lens focal length vs the objective focal length. Note that the objective magnification may differ from the actual

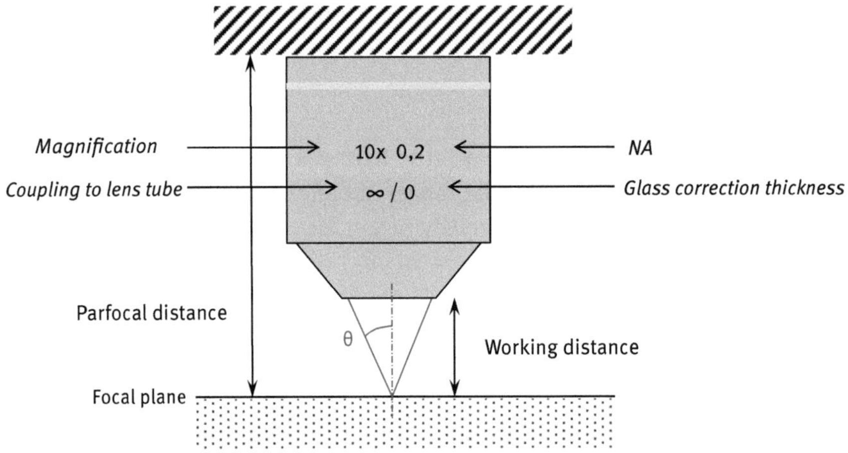

Figure 2.9: Objective keywords.

instrument magnification because of the use of a different camera lens. The base-side color ring is an indication of the magnification.
- *Working distance (WD)*: the distance between the front lens apex and the focal plane.
- *Parfocal distance*: the distance between the base of the objective and the focal plane. For a given microscope brand this value is constant.
- *Numerical aperture(NA)*: the angle from which light is collected by the objective (n: refractive index, θ half angle of the beam): $NA = n\sin(\theta)$

The objective can be chosen to accommodate a specific wavelength range, or a working distance, or higher NA (with immersion capabilities).

NA is very important because it defines the spatial resolution of the microscope (it's not the magnification) and in the end the Raman signal collection efficiency because of the ability to collect (or not) all emitted light. In that sense signal intensity for opaque samples is proportional to NA^2 (solid angle for collection). Quite often high magnification objectives have also high NA, and low magnification ones have low NA, so you will choose if you want a small or a large sampling volume depending on your experiment: for transparent bulk samples, low N.A. lens gives better sensitivity (because of the large sampling volume), whereas it's the opposite for opaque samples (because of the collection efficiency).

The spatial resolution in the case of a regular (nonconfocal) microscope is defined by Abbe formula for the diffraction limit, where the observed spot size diameter is 2 times the resolution:

$$D_{non\,confocal} = 1.22\,\frac{\lambda}{NA} \tag{2.2}$$

This is the resolution achievable on the video image, and differs from the resolution of the Raman confocal microscope.

2.4.4 Confocal microscope

Simply adding a microscope to a spectrometer assists in giving lateral (XY) spatial resolution, but does not give depth (Z) spatial discrimination because of the depth of focus. For this, confocal optics are required. There are several methods in use today, some truly confocal, others pseudo confocal, which work with varying success. For a true confocal design (which incorporates a real confocal pinhole aperture) micron depth resolution is possible, allowing individual layers of a sample to be discretely analyzed. The lateral resolution will also be increased.

As sometimes the substrate (or other layers) is fluorescent or gives Raman signal, the confocality will also help to reduce the background signals (Raman or fluorescence) coming from other parts of the sample and therefore will increase the Raman signal quality of the point of interest.

The principle is simple (see Figure 2.10): only light which is emitted from the focal plane will go through the confocal hole. Light emitted above or below will not, because the corresponding beam would be much larger than the confocal aperture. With the same principle, light emitted on the side of the focus point will be blocked on the side of the confocal aperture. That's why closing the confocal hole will also increase the lateral spatial resolution.

In summary, the confocal optics allows to define a "confocal volume" from where the light is collected, light coming from outside this volume being filtered out by the pinhole.

In the following example (Figure 2.11), a water inclusion (5 μm) in a quartz matrix with a gas bubble (1 μm), the semi-confocal then the confocal modes enable to get less interference from the matrix signal (quartz) and improve the signal for each fluid: water in semi-confocal mode, CO_2 gas in confocal mode.

2.4.5 Spatial resolution of Raman confocal microscopes

The ultimate spatial resolution achievable with an optical microscope is limited by the diffraction. In the specific case of a Raman confocal microscope, the main parts to consider while we deal with spatial resolution are the laser source, the microscope objective and the confocal hole aperture.

Concerning the laser source, the wavelength (λ) and the beam quality are the two parameters that can influence the spatial resolution.

Generally speaking the lower the wavelength, the better the spatial resolution. The beam quality of a laser can be defined by the M^2 factor, which represents how

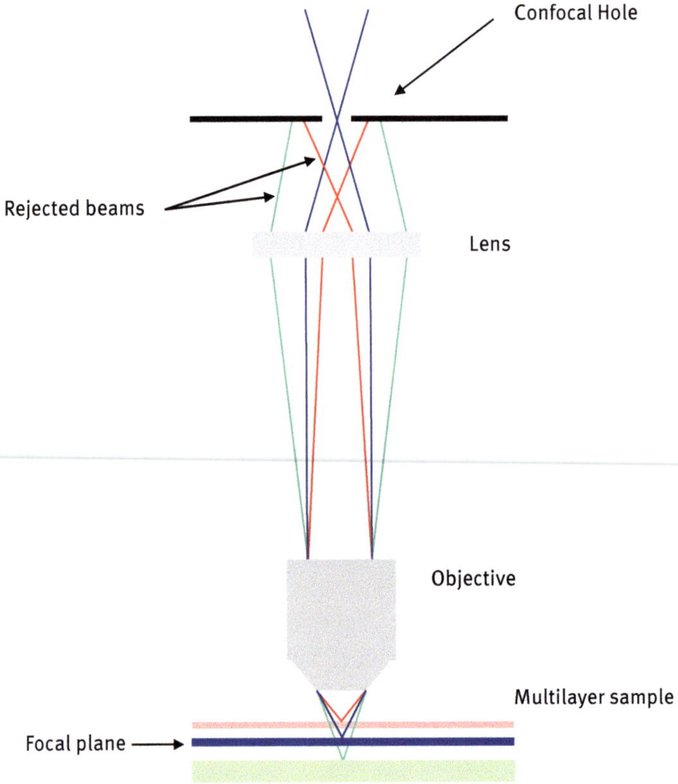

Figure 2.10: Principle of confocal microscope.

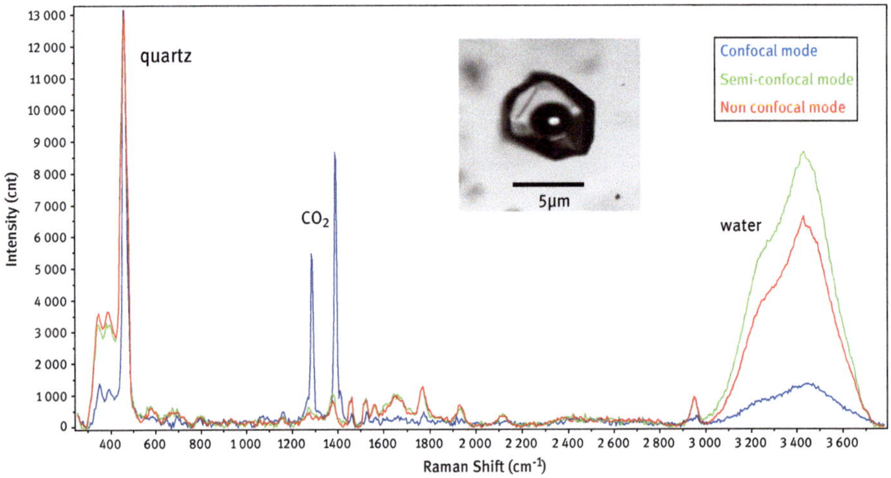

Figure 2.11: Effect of confocality to discriminate inclusions inside a matrix. All spectra were recorded at the same point, only the confocal hole size was changed.

tightly a laser beam can be focused. The best possible beam quality achievable is a M^2 equal to 1, which corresponds to a perfect Gaussian beam.

In the case of a confocal microscope, the microscope objective is one of the most important elements in determining the spatial resolution of the system. The numerical aperture (NA) of the objective is of great importance in confocal microscopy. The spatial resolution of the confocal microscope is directly dependent upon this value; the higher the numerical aperture, the better the lateral resolution.

It must be noted that the NA is the "effective NA", i.e., the effective light cone coming out of the microscope objective. It means that the laser beam size must match the entrance pupil size of the objective (this is called "pupil matching"). If the pupil is underfilled, the effective NA will be smaller and the calculations will be wrong.

Finally the confocal hole aperture, and its size in the object space defines how confocal the system is (see Figure 2.20 for magnification calculation):

- If the confocal hole aperture in the object space is higher than the spot size, then the spatial resolution is given by the formulas:

$$FWHM_{X,Y} = \frac{0.59\lambda M^2}{NA} \quad (2.3)$$

$$FWHM_Z = \frac{0.88\lambda M^2}{n - \sqrt{n^2 - NA^2}} \quad (2.4)$$

- If the confocal hole aperture in the object space is much smaller than the spot size (typically PH < 0.25FWHMgauss) then the lateral resolution is given by the formulas:

$$FWHM_{X,Y} = \frac{0.37\lambda M^2}{NA} \quad (2.5)$$

$$FWHM_Z = \frac{0.64\lambda M^2}{n - \sqrt{n^2 - NA^2}} \quad (2.6)$$

λ *= Laser wavelength,* ***NA*** *= Numerical aperture,* ***M²*** *= Quality factor of the laser,* ***n*** *= the refraction index of the medium (n = 1 for dry objective)*

In reality, if one select a confocal hole much smaller than the spot size, the Raman signal will be too weak and the noise will ruin the resolution, so usually (eqs. 2.3 and 2.4) are used.

From these equations it can be seen that lower wavelength lasers offer high spatial resolution (e. g. a blue laser at 488 nm will have a smaller spot size than an infrared laser at 785 nm if the same objective is used), as do high NA objectives (e. g. a 0.90/100x objective will give a smaller spot than a 0.55/50x objective).

Roughly speaking, because NA is usually close to 1, it can be considered that the spatial resolution is around two thirds of the laser wavelength.

The previous equations are slightly different from the usual equation for the Limit of Diffraction used in light microscopy (see eq. 2.2), because of the Gaussian profile of the excitation beam and because of the confocal hole.

The way to measure such spatial resolution and some results are shown at the Section 2.9.2.

Note: Be careful of the effect of depth penetration when using high NA objective to measure inside transparent samples: because of spherical aberration, it is necessary to use objective with a "glass thickness correction" (adjustable or not). This is very well-known when measuring a sample under a glass coverslip, but should not be forgotten when measuring below the surface of the sample. As shown on Figure 2.12, the rays with high incident angle will not focus at the same point inside the sample than small angle rays, because of light refraction at the sample's surface. If the sample is moved by dz, the central rays (with low angle) will focus at dz inside the sample, but the outer rays (large angle) will focus deeper at dz', degrading the total axial resolution

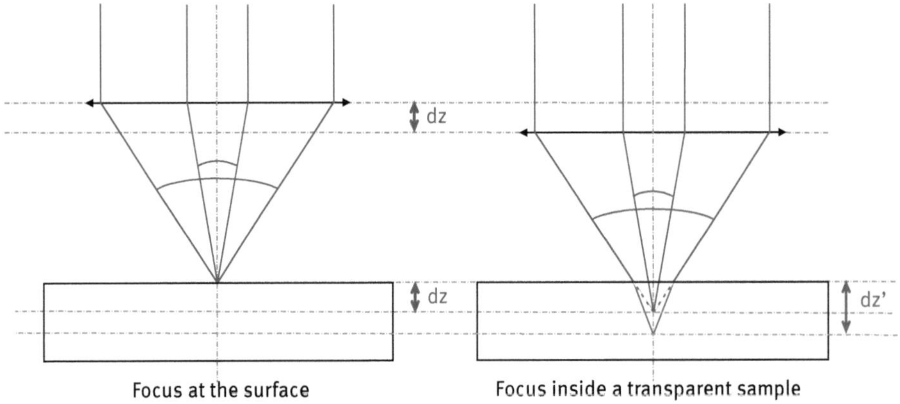

Figure 2.12: Effect of angle when focusing inside a transparent sample.

2.5 Filtering

2.5.1 Rayleigh filter (edge or notch)

The Rayleigh optical filter(s) are placed in the Raman beam path, and are used to selectively reflect the laser line and block the Rayleigh scattering (i.e., the laser wavelength), whilst allowing the Raman scattered light through to the spectrometer and to the detector. Remember that the Raman effect is approximately 1 million times lower than standard Rayleigh scattering, which means high efficiency filters must be used.

Usually the blocking performance is 10^6 and the transmission >90 %. Each laser wavelength requires an individual filter. There are two main types of filter used (see Figure 2.13), both of which can be used without user intervention or optimization:

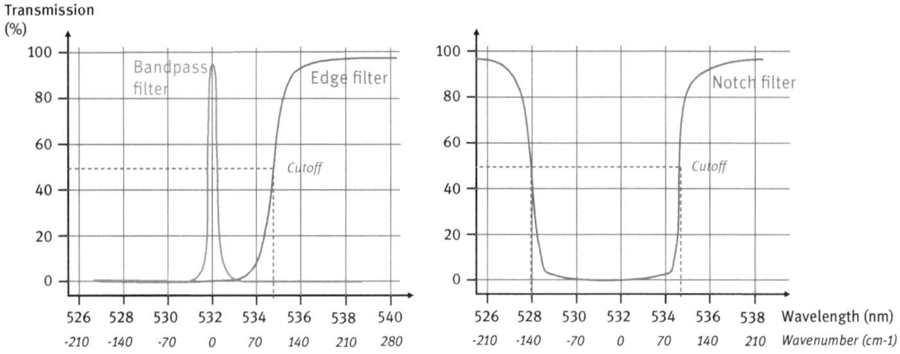

Figure 2.13: Principle of Bandpass (laser line filter), Edge and Notch filters, for a 532 nm excitation.

- **Edge**

An edge filter is a long pass optical filter which reflects all wavelengths up to a certain point, and then transmits with high efficiency all wavelengths above this point. As an example, a 532 nm edge filter will reflect all light up to 534 nm or above (e. g. 70 cm^{-1} or above) including the laser emission. Above 534 nm it will transmit light, allowing detection of the Raman spectrum (stretching from 70 cm^{-1} up to 3500 cm^{-1} or above). The edge filters have an ultra steep edge between the reflecting and transmitting spectral region, and offer excellent blocking of the laser line. Because the edge filter is made by dielectric coating technology, it is environmentally stable with a near infinite lifetime.

- **Notch**

A notch filter has a sharp, single rejection band which corresponds to a specific laser wavelength. Typically the rejection band will be a few nanometers wide (corresponding to a few hundred cm^{-1}). As an example, for a 532 nm laser a notch filter will be chosen with a central band at 532 nm. The laser line will be reflected, but the Raman spectrum above will be transmitted. The advantage of the notch is that it allows measurements to be made for both the Stokes and anti-Stokes Raman scattering, which is useful for certain specialised measurements. Notch filters are also manufactured these days using dielectric coating technology.

- **Holographic notch**

This type of Notch filter was used until the development of dielectric Edge and Notch filters. Unlike the dielectric filter, the holographic notch does have a finite lifetime, and will degrade with time (after a few years), that's why this technology has been abandoned.

- **Cutoff**: the wavenumber value from which it is possible to measure a Raman spectrum. Usually it is specified to be the 50 % transmission value, but actually one can measure from 10 or 20 %.

Remark: Tilting the filters: If you tilt a dielectric filter, you will change slightly its properties. The basic rule to remember is "a blue shift", which means that the transmission curves will shift to the lower wavelengths. This can be used to lower the cutoff of the filter, but at the same time, the blocking efficiency of the filter will decrease.

2.5.2 Triple monochromator: a tunable filtering spectrometer

With a triple monochromator instrument, it is possible to work in a "double subtractive, single spectrograph" mode. The first two monochromators are used as an integral pair, which firstly disperse both the Rayleigh (laser) and Raman scattered light, then physically block the Rayleigh, and then recombine the light. The third spectrometer then acts in the normal way to disperse the Raman scattered light with subsequent detection.

The advantage of using a triple monochromator in this way is that the laser filter is infinitely variable, and the one instrument can be used to work with any number of laser sources. In addition, the filtering performance is excellent, and allows Raman analysis down to 4–5 cm^{-1}.

However, such an instrument does require rather more expertise to operate in comparison with the more standard filter based single monochromator systems, and because of the number of optical elements involved it is much less efficient, requiring much longer integration times – as such it would rarely be considered for routine analysis.

2.5.3 VBG filters for ultra low frequency (ULF) measures

Recently have appeared a new kind of filters called VBG "Volume Bragg Gratings". These filters can work either in reflection or in transmission, and can be optimized as bandpass filters or notch filters. They have the advantage of being extremely narrow, their bandwidth being usually around 10 cm^{-1}. When cautiously integrated and adjusted into a Raman spectrometer, it is possible to reach ±5 cm^{-1} or slightly better (depending on the wavelength), on a regular "single" monochromator: the

Ultra Low Frequency region (<20 cm^{-1}) is not anymore limited to the use of the more complex triple monochromators (see an example in Figure 2.14). The transmission of the VBG filters is a bit lower than the dielectric edge or notch filter, but still much higher than the transmission of triple monochromator. Their main drawback is that due to the photoinscription process, they cannot be used at all in UV and are less efficient in the violet region. Please note that the overall performance in the low frequency region also depends on the laser spectral quality (especially spectral purity).

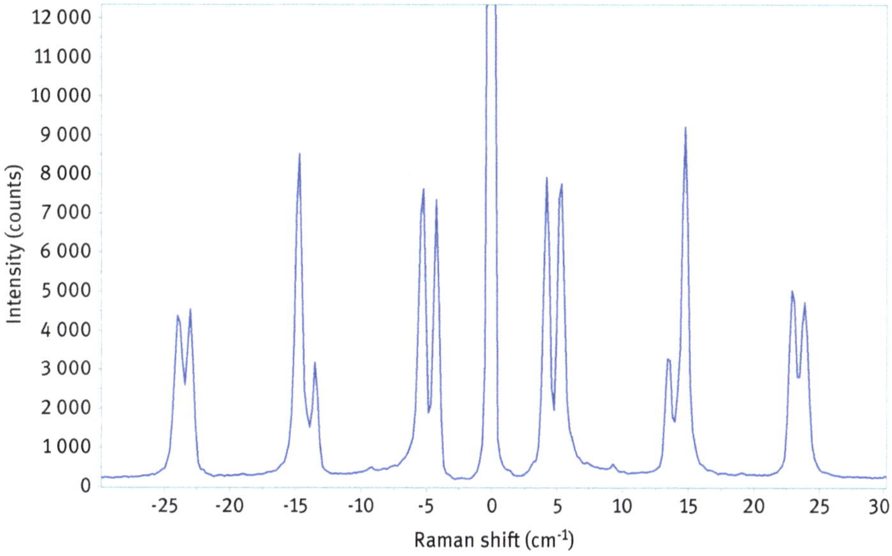

Figure 2.14: Ultra Low Frequency spectrum of SiGe superlattice, recorded with VBG filters (532 nm excitation, 3000 g/mm grating, 800 mm focal length) and showing bands down to ±3.5 cm^{-1}.

Table 2.6 summerizes the properties of the different filters.

Table 2.6: Summary of filters efficiency.

	Dielectric Edge	**Dielectric Notch**	**Triple monochromator in substractive mode**	**VBG**
Antistokes	N/A	−200 cm^{-1}	−5 cm^{-1}	−5 cm^{-1}
Stokes	50cm^{-1}	+100 cm^{-1}	+5 cm^{-1}	+5 cm^{-1}
Transmission	90 %	90 %	10 % to 20 %	70 %
Blocking wavelength	Fixed	Fixed	Tunable	Fixed

2.6 Spectrometer

2.6.1 Spectrometer principle

The spectrometer is the part of the Raman instrument that will analyze the scattered light and construct the spectrum. Here we will only talk about the dispersive spectrometers, that are the most commonly used.

Note that strictly speaking a monochromator is a device that will select only one wavelength. It shares the same design as the spectrometer.

To measure a spectrum, on can use a monochromator with a monochannel detector and adjust its wavelength by turning the grating, or use a CCD array to measure a portion of the spectrum in one time. One can also scan the grating of the spectrometer to increase the range of the detected spectrum.

Here is the principle of a dispersive spectrometer (from Figure 2.15):

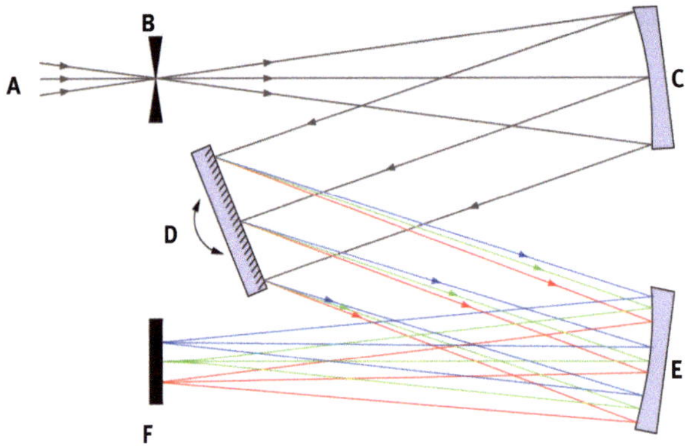

Figure 2.15: Czerney-Turner spectrometer.

The light (A) enters the spectrometer through the entrance slit B, is collimated by a first mirror C (or lens).

The grating D disperse the light (formula see)

Then a 2nd mirror/lens E focuses the light on the CCD (F).

In a few words, we can see the functions of each element as:

- The grating converts the wavelength into an angle
- The mirror/lens converts the angle into position on the focal plane

The spectrometer configuration shown just above is called "Czerney-Turner", and has the advantage to be tunable (by rotating the grating) and easily adjusted or modified (by changing the grating). The use of mirrors instead of lenses prevent chromatic aberrations.

Some other system work with concave gratings but they have fixed design. The concave surface images the entrance slit to the CCD, while the grooves on the surface of the mirror disperse the light.

2.6.2 Gratings

The grating is the core element of the spectrometer. It is a diffractive element composed by a series of parallel "lines" that will reflect the light, each line (or groove) generating a different optical path length for the light beam. The combination of these differential light travel distances generates interference patterns where the constructive interferences are called "orders of diffraction".

This results in the following equation, known as the grating formula:

$$\sin\alpha + \sin\beta = k.n.\lambda \tag{2.7}$$

where:

α: incoming angle

β: output angle

k: order of diffraction

n: line (groove) density, in groove/mm (g/mm)

The gratings will be most efficient in the first order of diffraction (k = 1). The zero order corresponds to regular reflection of light (see Figure 2.16). One can see from the equation that at the maximum sin a + sin b = 2, so not all wavelength are accessible to a grating: the theoretical limit wavelength for a grating of groove density n is lambda = 2/n.

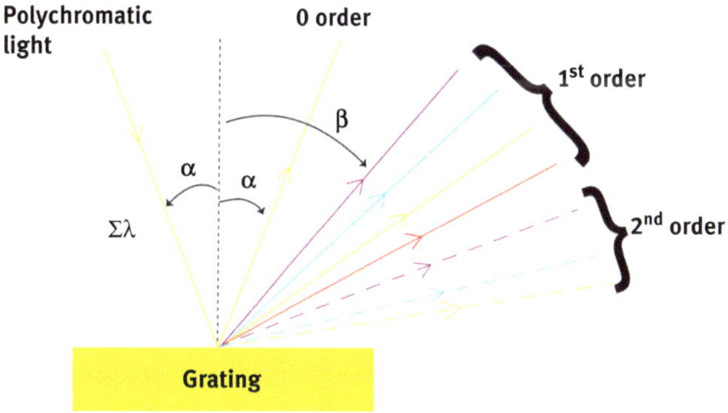

Figure 2.16: Grating operation.

For instance, a 1800 g/mm can be used until 2/1800 = 1100 nm (in reality the limit is more like 900 nm in a real spectrometer geometry).

By changing the profile of the lines, it is possible to boost the efficiency of the grating at a specific wavelength range. This operation is usually done by tilting the angle of the line, which is called the blaze angle (Figure 2.17). Such a grating is named “blazed at wavelength [. . .]” or just written “@[. . .]”

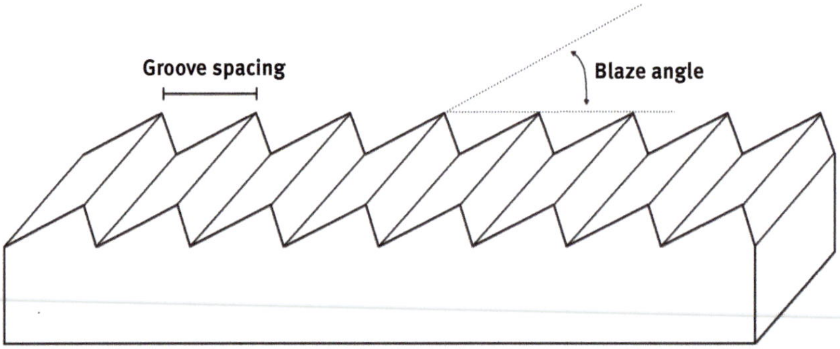

Figure 2.17: Grating grooves profile.

As a reminder, there is no mystery: the broader the spectral range, the lower the efficiency.

So when you select a grating, take care also of the blaze wavelength, which can yields to significant differences in efficiency, as shown in Figure 2.18.

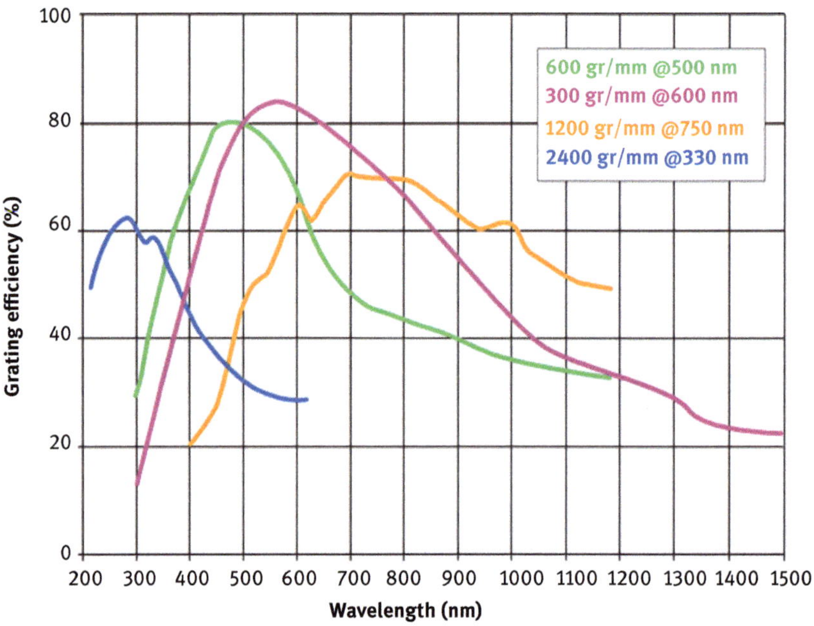

Figure 2.18: Different grating efficiency curves, depending on groove density and blaze wavelength.

2.6.3 Grating drive, linearity, calibration

Spectrometer can be fixed or scanning ones. In the first category, the grating is fixed, which means that the wavelength range that is selected and focused on the detector is always the same. On the contrary in a scanning spectrometer, the grating can rotate and so the wavelength range can be shifted.

As a small rule, the range is constant in nm for a given grating (but not in wavenumber).

To rotate the grating, the most straightforward way is to place the grating on a rotary drive.

On older design instruments, a linear movement (from a screw with a nut) is converted into the rotation by a mechanical sine function (so that the wavelength is proportional to the displacement). It was invented when the computers didn't know the sine function, but is still used because of its proven accuracy and reliability over the years.

To ensure the correct performance of a spectrometer, some corrections are necessary:
- Linearity (or drive error) consists of having the correct wavelength at the center of the detector for each angular position of the grating (because the rotating wheel is not perfect for instance).
- Calibration is the accurate calculation of the wavelength on every pixel of the CCD (based on exact grating grooves density, angle, focal length ...), over the whole wavelength range.

2.6.4 Slit

A monochromator builds the image of its entrance aperture to its exit. As the dispersion occurs in a plane perpendicular to grating groves, the image height at the monochromator exit carry no spectral information and contributes only in detected spectrum intensity. That's why the entrance aperture is a slit, which width defines the spectral resolution.

Usually, 3 types of entrance slit can be found: fixed slit (only one width), slit with several different sizes and slit with continuously variable size.

Why change the slit size or more precisely, width? The slit is located at the image plane of the confocal hole and therefore of the sample. Figure 2.19 shows the image of the laser spot at the sample on the slit plane. The choice of the slit width is a compromise between spectral resolution and quantity of detected light:
- If the slit width is reduced it will increase the spectral resolution but reduce the light transmission.
- On the opposite, opening the slit will increase the light throughput at the cost of a worse spectral resolution.

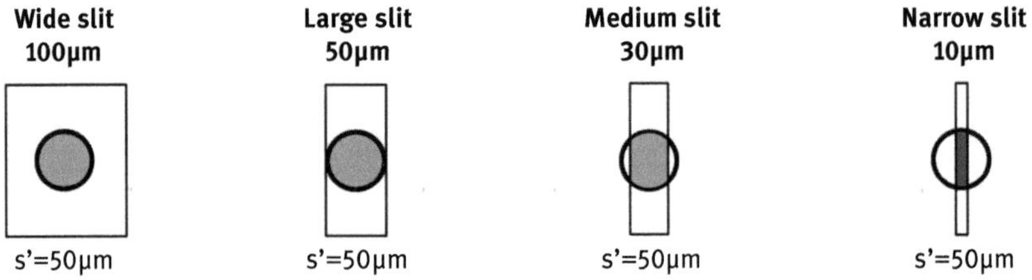

Figure 2.19: Slit size vs spot size image (s').

Sometimes, in some instruments, the confocal hole and the spectrometer entrance slit are the same thing. In this case spectral resolution (defined by the slit width) and spatial resolution (defined by the hole diameter) cannot be separated. In order for the system to remain truly confocal, it should be a hole. If it's a slit, then it's called "pseudo confocal" because in 1 direction there is no spatial filtering, and the z resolution will actually be poor.

2.6.5 Optics and spectrometer coupling

In order to be the most efficient, the spectrometer must be optically adapted to the whole device setup, if not, the beam cross-section will be larger than the spectrometer optics, or the entrance slit/hole will strongly limit the light entering the spectrometer. This is called "aperture matching" of the spectrometer.

For this, the f-number of spectrometer, defined by the size of its optics and its focal length (see eq. below), must match the f-number of the beam focused on the slit. The f-number can be adapted by using lenses, but because of energy conservation principle, this will affect (usually in the wrong way) the size of the image, and therefore the spectral resolution or the light throughput.

This aperture matching is done by every Raman spectrometer manufacturer and is optimized for different conditions and hypothesis.

The relationship between f-number (f#) and NA is:

$$NA = n.\sin\alpha \tag{2.8}$$

We call "f-number" N:

$$N = \frac{f}{D} \approx \frac{1}{2NA} \tag{2.9}$$

The resulting f-number (N) = 10 is written as f/10

For instance, in the case of a direct coupling between the microscope and the spectrometer, a low NA spectrometer (f/10 or above) is usually enough because the magnification between the slit and the objective is big.

On the contrary in the case of a fiber coupled spectrometer a high NA is better (f/2), because the NA of an optical fiber is usually around 0.22, which is f/2.3. But it's possible to add an adaptor before the entrance slit.

In the example of Figure 2.20 the magnification between the sample and the slit is $\left[\frac{f'}{f}\right] = 50$, the NA of the objective is 0.9 which is f/0.55, so the spectrometer aperture is f/28 or $NA_S = 0.018$:

$$N = \frac{f}{D} \approx \frac{1}{2NA} = \frac{1}{2 \times 0.9} = 0.55 \text{ and } N_S = \frac{f_S}{D_S} = \frac{f'}{D} = \left[\frac{f'}{f}\right] N = 28$$

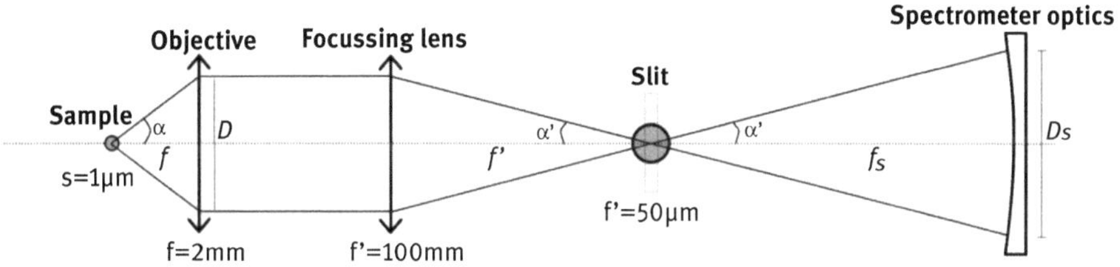

Figure 2.20: Coupling optics.

2.6.6 Spectral resolution/dispersion

Spectral resolution in a dispersive Raman spectrometer is determined by five factors. In the discussions below, the effect of each factor is considered under the assumption that all other factors remain unchanged. In real life all of these factors can exist in many varied permutations, which makes direct comparison of a system's performance and capabilities difficult.

Dispersion or resolution?

Spectral dispersion is calculated in nm/pixel or cm^{-1}/pixel. It is a geometric assumption of the size of a pixel in nm (or cm^{-1}). It will tell how many pixels there are in a given spectral range. It depends on the grating and the spectrometer focal length. In short this is the "wavelength scale" of the spectrum.

Spectral resolution is measured in nm or cm^{-1}. It is a property of the full spectrometer that qualifies how well the peaks are separated from each other. In short it is the "optical sharpness" of the spectrum. It depends on the spectral dispersion, the slit size and the optics quality.

High dispersion does not always mean high resolution, because of the slit size and the optical aberrations of the spectrometer.

- **Spectrometer focal length** the longer the focal length (e. g. the distance between the dispersing grating and detector) of the spectrometer the higher the spectral dispersion (and usually spectral resolution) – see Figure 2.21.

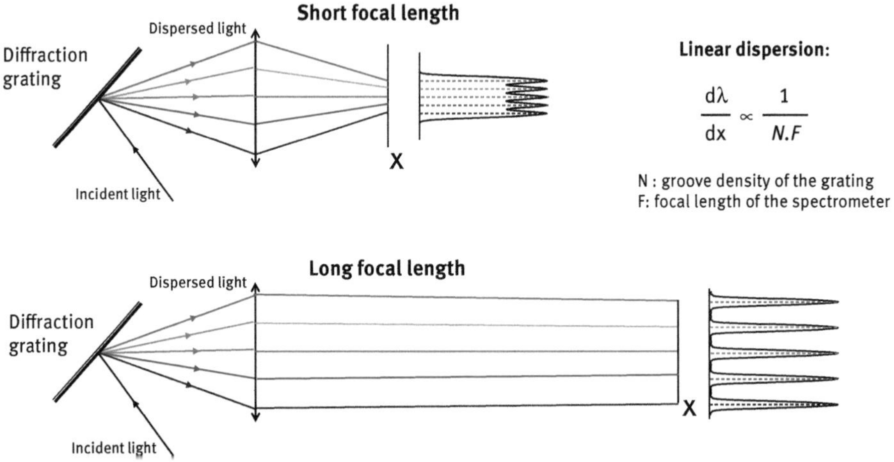

Figure 2.21: Influence of focal length on spectral dispersion.

Typical Raman spectrometers have focal lengths ranging from 200 mm (for low/medium resolution) through to 800 mm and higher (for high resolution). It is sometimes forgotten that a long focal length spectrometer is not limited to high resolution work only – with suitable choice of grating (see below), a high resolution spectrometer can be run in a low resolution mode, while the opposite is not true.

- **Diffraction grating** – the higher the groove density of the grating (typically measured as number of grooves per millimeter or gr/mm), the higher the spectral dispersion – see Figure 2.22. Typical gratings used for Raman vary from 150 gr/mm (low dispersion) through to 2400 gr/mm (high dispersion) or even higher. The use of higher groove density gratings cannot be applied *ad infinitum* to increase spectral dispersion, since they will have fixed practical and physical limits linked with the spectrometer itself (see grating equation). Thus gratings provide an initial way to improve dispersion, but once their limit is reached it is necessary to move to a longer focal spectrometer.

  Ex: the dispersion of a 200 mm focal length spectrometer equipped with 2400 g/mm is equivalent to 400 mm focal with 1200 g/mm or 800 mm with 600 g/mm.
- **Laser wavelength** – the dispersing power of a grating/spectrometer pair can usually be considered constant in terms of wavelength. However, Raman spectra use an energy related unit (Raman shift, or wavenumber, cm^{-1}) which means that the spectral resolution decreases as the laser excitation is changed from infra-red to visible to ultra-violet wavelengths (as deducted from equation 2.10 and Table 2.7). As an example, if a 600 gr/mm grating is used with an infra-red laser, a 1200 gr/mm or 1800 gr/mm will be required with a green laser to achieve a similar resolution.

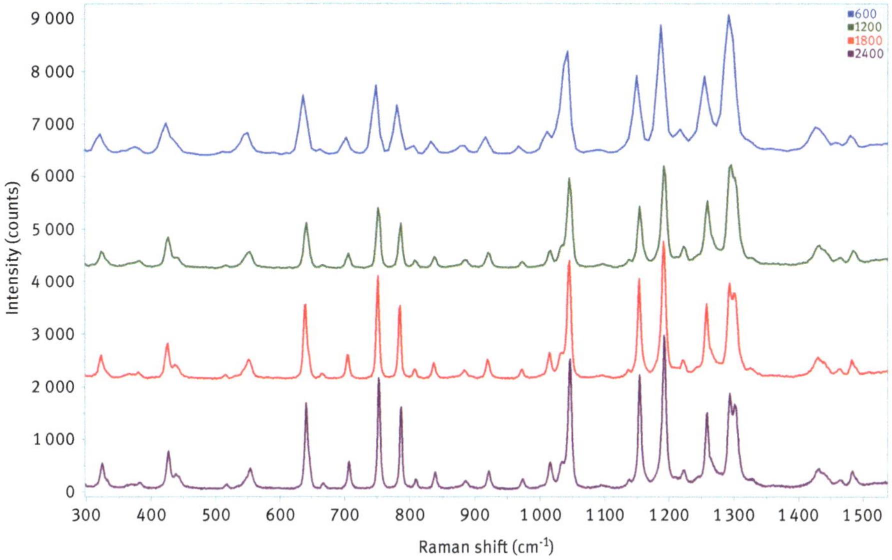

Figure 2.22: Influence of the grating on the spectral resolution of a tablet's spectrum (532 nm excitation). See especially the differences at 430, 1050 and 1300 cm^{-1}.

Table 2.7: Equivalence between wavenumbers and nanometers.

Wavelength	1 nm represents
325 nm	94 cm^{-1}
532 nm	35 cm^{-1}
785 nm	16 cm^{-1}

The conversion formula between nanometer and wavenumber is :

$$\nu_{raman}\left(cm^{-1}\right) = \left(\frac{1}{\lambda_{laser}(nm)} - \frac{1}{\lambda_{raman}(nm)}\right) \times 10^7 \tag{2.10}$$

– **Detector** – most systems have a single detector, so practically the user does not have control of this factor. However, it should be noted that different detectors can be configured with different pixel sizes. Basically, the smaller the pixel, the higher the achievable spectral resolution. But do not mix peak definition (the number of pixels in one peak) and spectral resolution. You will get a better resolution with small pixels only if the slit is also smaller (which can lead to smaller signal).

Note that, in the opposite way, it is also possible to bin the detector (combining adjacent pixels together) which artificially increases the pixel size to have higher signal and lower resolution.

- **Slit size –** The slit size plays a crucial role in the spectral resolution, as explained at Section 2.6.4. But it will have no effect on the "spectral dispersion".

  If the slit is totally illuminated, a slit of 100 µm width used with a 1800 gr/mm grating will give the same spectral resolution than a 50 µm slit used with a 900 gr/mm grating. But the spectral "window" observed with the 1800 gr/mm grating will be half of the one with 900 gr/mm.

2.7 Detector

2.7.1 What is a CCD detector?

A CCD (Charge Coupled Device) is a silicon based multichannel array detector of UV, visible and near-infra light. They are used for Raman spectroscopy because they are extremely sensitive to light (and thus suitable for analysis of the inherently weak Raman signal), and allow multichannel operation (which means that the entire Raman spectrum can be detected in a single acquisition). CCDs are widely used in today's technologies, not least as the sensors in digital cameras, but versions for scientific spectroscopy are of a considerably higher grade to give the best possible sensitivity, uniformity and noise characteristics.

CCD detectors are typically one dimensional (linear) or two dimensional (area) arrays of thousands or millions of individual detector elements (also known as pixels). Each element interacts with light to build up a charge – the brighter the light, and/or the longer the interaction, the more charge is registered. At the end of the measurement, read out electronics pull the charge from the elements, at which point each individual charge reading is measured in the ADC (Analog to Digital Converter) - see the read-out details in Figure 2.23.

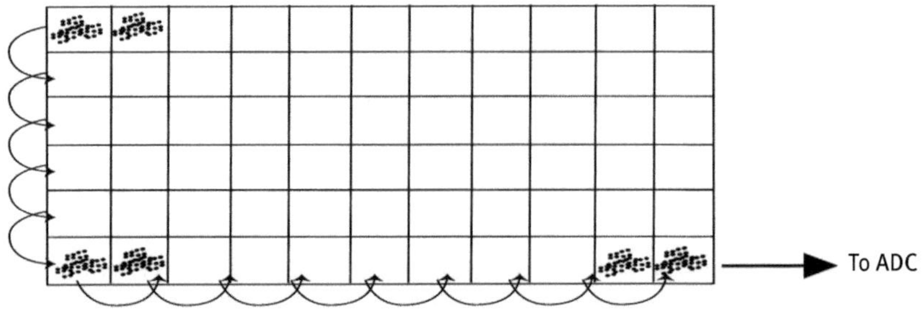

Figure 2.23: CCD readout principle: After exposure, the charges at each pixel are transferred line to line down to the readout register where they are then horizontally transferred towards the digital converter.

In a typical Raman spectrometer, the Raman scattered light is dispersed using the diffraction grating, and this dispersed light is then projected onto the long axis of the CCD array. The first element will detect light from the low cm^{-1} edge of the spectrum, the second element will detect light from the next spectral position, and so on ... the last element will detect light from the high cm^{-1} edge of the spectrum.

CCDs require some degree of cooling to make them suitable for high grade spectroscopy. Typically this is done using either Peltier cooling (suitable for temperatures down to −50 °C, −70 °C or even −90 °C with additional water cooling) or liquid nitrogen cryogenic cooling (this is more rare and limited to the most demanding applications).

2.7.2 Quantum efficiency

Quantum Efficiency (QE) is a characteristic of a CCD detector, defined as the ratio of conversion from photons into electrons. The different types of CCD have different QE curves over wavelength, as shown in Figure 2.24. The CCD being Silicon-based sensors usually covers the UV and the visible range, up to the Near Infrared.

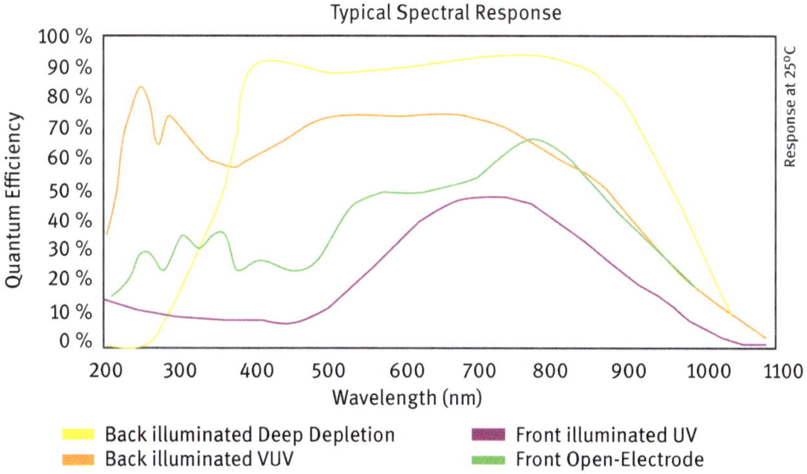

Figure 2.24: Comparison of different CCD quantum efficiencies.

2.7.3 Noise

There are three sources of noise to consider on the detector:

– Readout noise (e): induced by the electronics and the analog to digital conversion. It depends on the read-out speed (the faster, the more noise).

- Thermal/dark noise (e/s): from the sensor itself, and linked to integration time. When doing fast imaging (<100 ms), this can be neglectable.
- Shot noise (photon/s): proportional to light intensity, present when there's high signal, which is usually not the case in Raman spectroscopy, but can occur if you have fluorescent background.

When you compare CCD, don't forget to look at the noise levels, because even if QE curves may look more interesting, depending on your application (and integration time) the noise level can ruin totally the benefit of higher QE (see for instance the comparison between CCD and IGA detector in Section 2.7.7).

2.7.4 Etaloning effect

In the case of Back-Illuminated (BI) CCD, there is an interference effect that occurs in the Silicon chip itself, cause by the reflections between the top surface and the focal plane: its is called "etaloning effect". This effect is seen in Red or NIR range of detection, especially for the "BI-VIS" or "BI-UV" chip, and can be seen as a much higher noise in these spectral domains. The Figure 2.25 shows this effect on the spectrum.

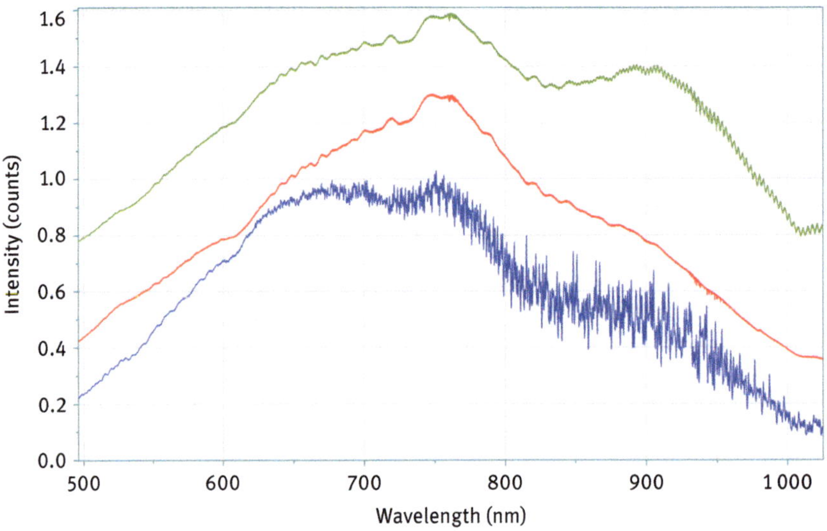

Figure 2.25: White light response of three different CCD: Open electrode (red) without etaloning, BI-UV (blue) with a strong etaloning from 650 nm, BI-DD (green) with "reduced etaloning" treatment (etaloning is present from 850 nm).

Over the last years manufacturers have developed "reduced etaloning" chip with a special coating or treatment made on the first surface, to reduce the reflectivity and thus limit the interferences. The efficiency is variable depending on the detector, but usually it will always remain visible, from 1 % to 10 % relative intensity. Without the "reduced etaloning" feature, the level would be around 20 % to 40 %!

2.7.5 Types of CCD – comparison table

The difference between the different types of CCD detectors is summarized in the following table (Table 2.8):

Table 2.8: Comparison of different CCD detectors.

CCD chip type	Use	QE	Noise	Spectral range	Comment
OE – Open Electrode	Generic purpose	medium	excellent (very low noise)	Broadband UV-NIR	
FI-VIS, FI-UV: Front Illuminated	No interest for Raman	Medium	excellent	Broadband UV-NIR	Smaller pixels available
BI-UV: Back Illuminated UV	UV	high	medium	UV or blue	Strong etaloning in visible and NIR range
BI-VIS: Back Illuminated visible	No interest for Raman	high	medium	VIS-NIR	Strong etaloning in visible and NIR range
BI-DD: Back Illuminated Deep Depletion	Short integration time, no background	high	medium	VIS-NIR	Etaloning in NIR
EMCCD	Fast imaging	Depends on type	medium (depends on type)	Depends on type	See next section

2.7.6 EMCCD

An Electron Multiplying CCD (EMCCD) is a special type of CCD detector, which enhance the spectrum quality when extremely low signal levels are present. After the photons have been converted into electrons in the chip, the electrons are analogically multiplied with an adjustable gain factor (from 1 to ~1000).

The key benefit of an EMCCD is that the amplification occurs before readout of the signal, which means that the signal is not readout noise limited. In other words, through amplification the signal is raised well above the noise floor which is largely determined by the noise of the readout electronics (pre-amplifier and A/D convertor).

2.7.6.1 When to use EMCCD?

The benefits of EM gain are clearly obvious in fast Raman spectral imaging, where the necessary short integration times can often result in signals which are barely visible above the noise when measured with a conventional CCD.

Be careful, as a limitation, that there should be no background, because it would be amplified too and the Raman signal will still be mixed with the background.

Similarly to the different CCD, there are different types of EMCCD (with almost the same names): BIUV, BIDD, FI, FIUV, but there is no equivalent of the multipurpose OE CCD.

2.7.7 Infrared detectors

2.7.7.1 IGA

InGaAs detectors (or IGA) have sensitivity in the infrared (800 nm–1600 nm for the usual version).

But they have much higher dark noise level (typically 1000 times higher), which requires deeper cooling (usually cryogenic Liquid Nitrogen "LN" or "LN2").

Even if the QE looks good compared to a CCD in the NIR region, always use a CCD when it is still sensitive, because the signal-to-noise ratio will always be higher with a CCD (see Figure 2.26 below).

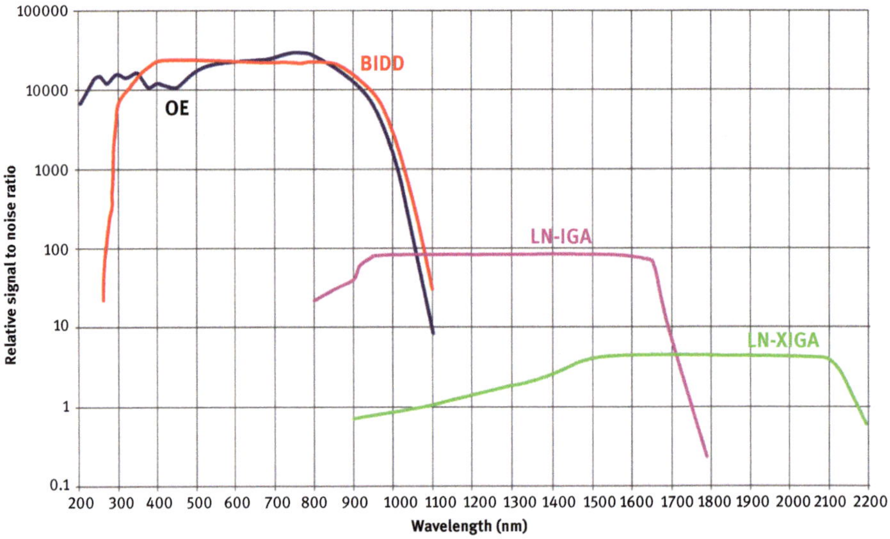

Figure 2.26: Relative efficiency of different detector types for 1s acquisition time (logarithmic scale).

2.7.7.2 X-IGA

Extended InGaAs (XIGA) detectors are named this way because their spectral range is extended towards infrared. Their sensitivity can reach 2.2 μm, but the drawback is a significant increase of the noise, compared to the regular IGA.

2.7.8 Monochannels detectors

Consider these as obsolete, for VIS or NIR range, and even if they are perhaps cheaper than a multichannel IGA for NIR, they show less performance, much more complex use and less robust.

2.8 Software, automation, electronics

All manufacturers have developed their own software to drive their Raman spectrometer and perform data acquisition and treatment, with lots of parameters, but more and more automation and ease of use.

Basically everything can be automated in the instrument, the most common features being: grating angle (to select appropriate range), grating selection, excitation wavelength selection (laser and associated filters), adjustment of laser power, size of confocal hole, entrance slit, activation of video, moving the sample ...

Most softwares have now integrated more and more data treatment functions, from the spectra correction (noise removal, background removal ...) to the analysis (peak fitting, normalization, modelling, multivariate analysis, database searching ...).

2.9 Quality assessment

There is currently no normalized way to measure all performances of a Raman microscope. There are some usual tests often provided.

2.9.1 S/N ratio

Several manufacturers use signal to noise evaluation by measuring a Polystyrene spectrum. Signal is the integral of the 1000 cm^{-1} peak and noise is measured in the spectrum zone with no peak - see Figure 2.27. Be careful that this value depends on a lot of parameters: laser, objective, dispersion, confocality ... and of course acquisition time.

A more historical way is to measure the 3rd order or the 4th order of the Si spectrum, but there is the same difficulty to quantify or to compare between instruments

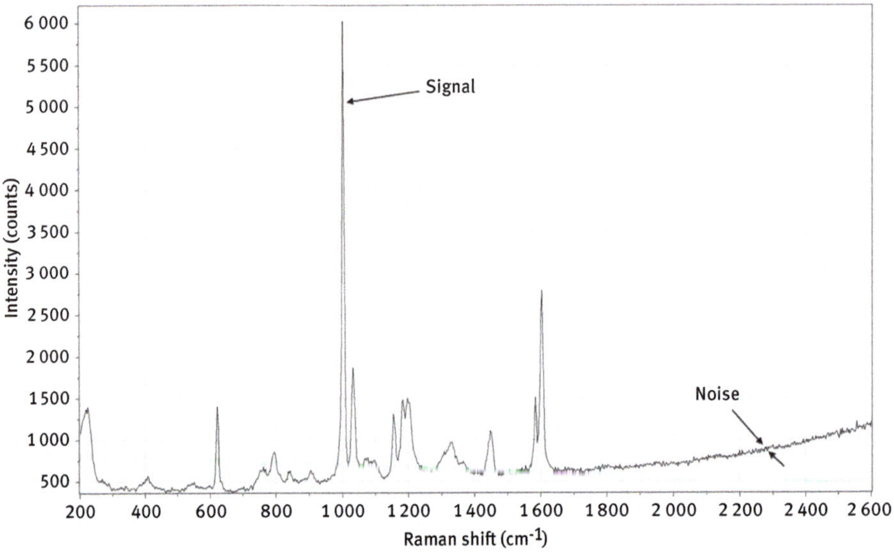

Figure 2.27: Signal to noise ratio measured using polystyrene spectrum.

because of the number of instrumental parameters involved. On the contrary to the polystyrene spectrum, it is more difficult to measure due to the very low relative intensity compared to the main Silicon peak (see Figure 2.28).

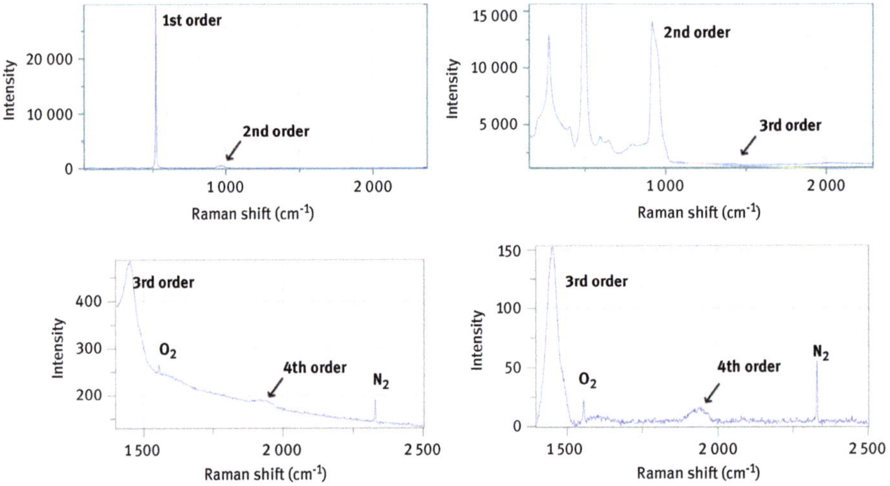

Figure 2.28: Silicon spectrum 1st, 2nd, 3rd and 4th order. Upper left with 0.5s acquisition. Upper right with 30s acquisition. Bottom left with 60s acquisition. Bottom right same after baseline substraction.

2.9.2 Spatial resolution

Spatial resolution (x-y) can be measured using patterned samples (like metal stripes on Silicon), a sharp transition (like metal deposed on silicon) or an ultrathin sample (like a carbon nano tube). Z resolution needs a nonpenetrating sample (Silicon for blue or green laser, Germanium for red or IR), or an ultrathin one (like suspended graphene monolayer).

The general principle is shown on the Figure 2.29: it consists in performing a 1D scan over an edge (step function) and calculating the first derivative. The FWHM of the resulting curve gives you the point spread function (PSF) which FWHM represents the spatial resolution. The alternative is to perform the scan over an "infinitely thin" sample, equivalent to a Dirac function, and the PSF is directly equal to the measured profile. In the case of an intermediate thickness, it must be taken into account to deconvolute the result.

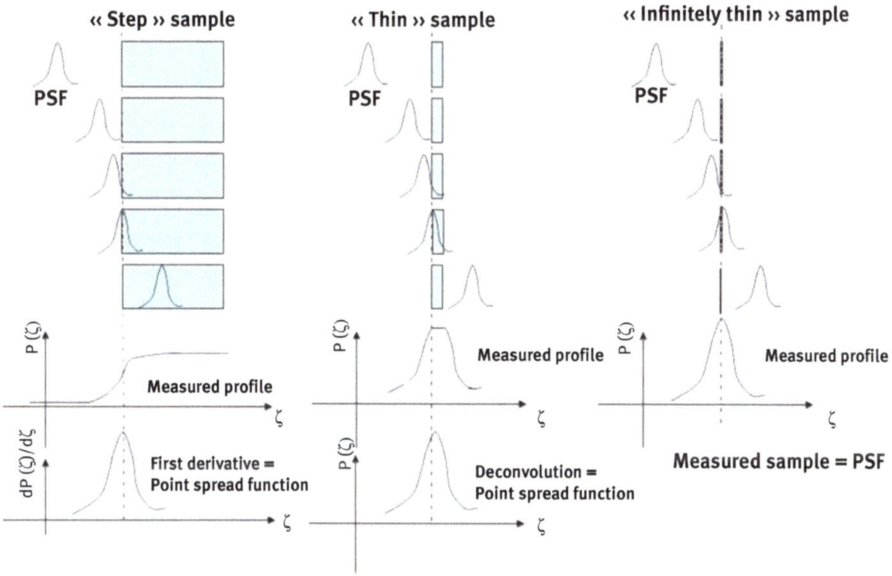

Figure 2.29: Spatial profiling of a sample used to measured either the axial or the lateral PSF of the confocal microscope.

Obviously, it is important to record such a profile with a sampling rate that respects the Shannon-Nyquist criteria (The sampling rate of a signal must be at least two times higher than the maximum frequency contained in the signal). In practice, the lower the step size, the better the definition of the profile.

In the following example (Figure 2.30) the lateral resolution has been measured on a 0.1–5 Si-Cu pattern (0.1 µm Si, periodicity 5 µm) sample. The measured lateral

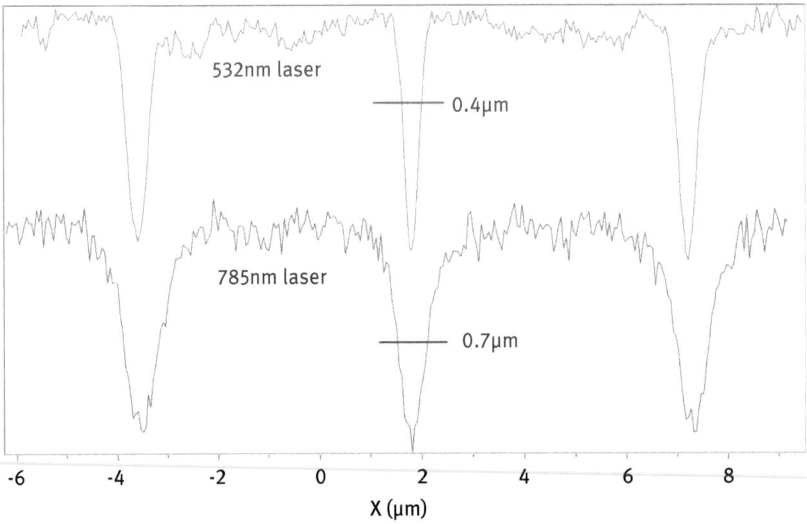

Figure 2.30: Lateral profile measured on a 0.1 µm patterned Si sample, with 532 nm and 785 nm excitation, 100x/0.9 objective.

profiles show lateral profiles having a FWHM as low as 0.4 µm for the 532 nm laser and 100X objective (NA 0.9). The resulting lateral resolution after deconvolution is as good as 0.3 µm which again is very close to the diffraction limit theoretical value for a perfect Gaussian beam (301 nm). Note the significant loss in resolution by using a NIR laser, because of the wavelength value and the M^2.

To measure depth resolution, we can use a Ge(111) sample having a very low depth penetration (20 nm at 488 nm, 85 nm at 633 nm). The following Z profile was done at 457 nm with a 100X objective on a Ge sample. Again, the diffraction limit theoretical value was nearly achieved in this example (Figure 2.31).

Remember that the achievable depth resolution will depend strongly on the laser wavelength, microscope objective, and sample structure.

2.10 Advanced

2.10.1 Raman imaging

Raman spectral imaging (or mapping) is a method for generating detailed chemical images based on a sample's Raman spectrum. A complete spectrum is acquired at every pixel of the image, and then interrogated to generate false colour images based on material composition and structure:

– Raman peak intensity yields images of material concentration and distribution

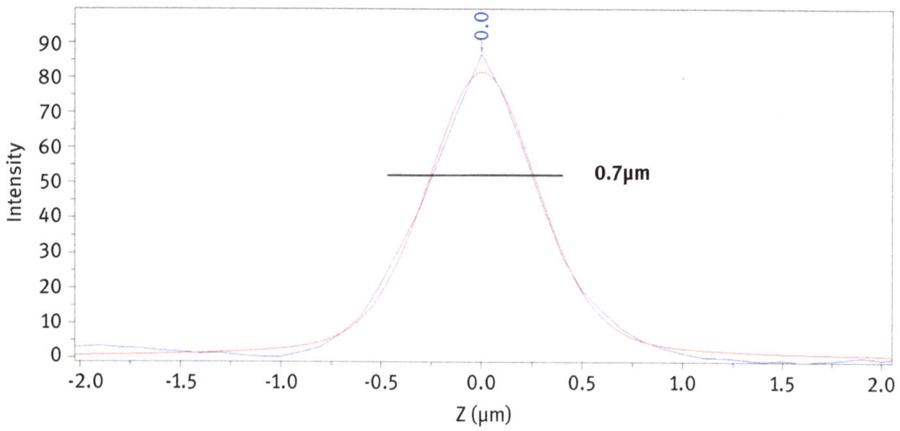

Figure 2.31: Depth resolution (derivative of the depth profile) measured on a Ge sample, with 457 nm excitation and a 100x/0.95 objective.

- Raman peak position yields images of molecular structure and phase, and material stress/strain
- Raman peak width yields images of crystallinity and phase
  Raman spectrum as a whole allows material identification (as in Figure 2.32)

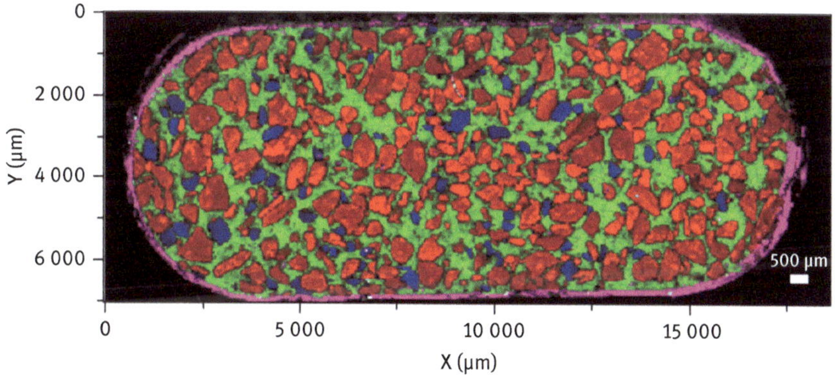

Figure 2.32: Raman spectral image of a whole sectioned pharmaceutical tablet. (red: aspirin, green: paracetamol, blue: caffeine, purple: coating).

A typical experiment uses sequential sample movement and spectrum acquisition, repeated hundreds, thousands or even millions of times, to collect data from the user defined image area.

Raman spectral images can be collected in two and three dimensions, to yield XY images, XZ and YZ slices, and XYZ datacubes.

Raman spectral imaging is an invaluable technique for scientists in many varied fields, since it allows chemical distribution to be viewed which is invisible by standard optical microscopy.

2.10.1.1 Sample scan

The first way to generate an image is to move the sample under the objective, using the microscope stage (X, Y and Z). Two types are used:

- motorized stages (xy), using stepper motors, sometimes with an encoding system for better accuracy. The Z motor is controlling the movement along the Z axis of the microscope.Their scanning range is several cm to hundreds of cm, with an accuracy around 1 µm and a step size of 100 nm or below.
- piezo stages, using piezoelectric actuators (xy or xyz). They are faster but with a scanning range usually much more limited (100 µm). They have higher accuracy and step size (in the range of the nm). They are also more expensive.

2.10.1.2 Laser Scan

Another option is to keep the sample steady, and to scan the measuring spot on the sample surface. This is typically done by a scanning mirror (or several) which is placed on the laser/Raman common path.

2.10.1.3 Line scan

Instead of scanning a point on the sample, a line can be used to measure simultaneously several points, by taking advantage of the slit height of the spectrometer.

Every point of the line is imaged on a point of the slit, and on the detector you measure all corresponding spectra “vertically”.

The drawback of this principle is to remove the confocal aspect.

2.10.1.4 Effect of spatial resolution

The confocal properties of the Raman microscope will directly affect the spatial resolution of the Raman images when dealing with small objects. The Figure 2.33 shows the effect on spatial resolution of the same sample by going from “not confocal” to “confocal” acquisition mode on the same instrument.

2.10.1.5 Fast imaging

Fast Raman spectral imaging is a technique for allowing Raman spectral images to be acquired with acquisition times down to less than a few ms/point, resulting in total measurement times of seconds or minutes, even for images comprising tens or hundreds of thousands of spectra.

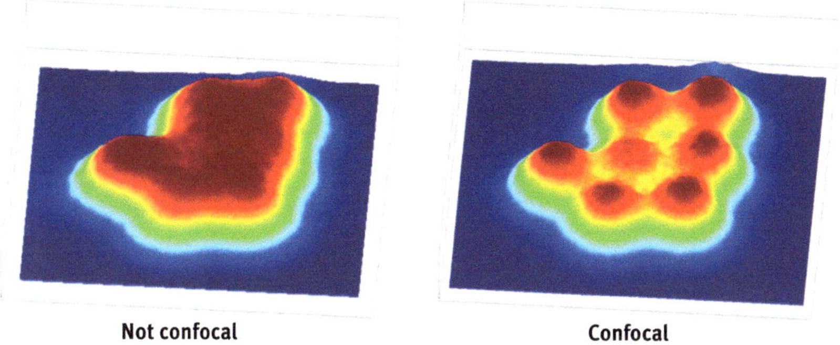

Figure 2.33: Not confocal and confocal Raman images of 1 µm polymer beads (3D rendering of 2D images with height as intensity).

Traditional Raman spectral imaging has always been limited by long acquisition times, but such methods do offer the ultimate sensitivity for materials with extremely low Raman scattering properties, and additionally allows high resolution, large spectral range measurements. Typical acquisition times for such maps can be in the order of 1s–10s per point (or longer), resulting in total measurement times in the order of hours or days.

Methods for Fast Raman spectral imaging vary, but typically they coordinate sample stage movement (or laser spot deflection) and detector readout to minimize "dead time" between pixels which occurs in standard point-by-point imaging experiments. They allow Raman spectral images to be acquired with acquisition times down to 1 ms/point or less, so that large area survey scans and detailed Raman spectral images can be completed in seconds or minutes, like for instance Figure 2.34.

Ultra-fast Raman spectral imaging is not suitable for every type of sample, and its efficacy will depend on the sample's inherent Raman intensity, and the required spectral quality necessary to create the image.

2.10.2 Autofocus/3D imaging

When adding a z-motor to the microscope, it's possible to reach different focus points during an experiment. This can be used to perform autofocus and 3D mapping (surfacic or volumetric).

2.10.2.1 Autofocus

Usually the sample is not flat. And even if it is flat is can be tilted versus the focal plane.

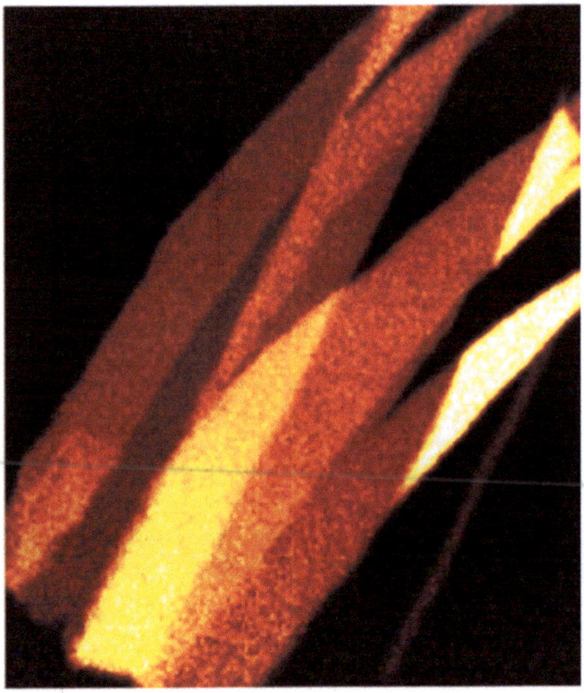

Figure 2.34: Various graphene layers – 28,080 spectra recorded in 71s (step 0.5 µm).

Depending on the sample scale and the size of your image, you may want to ensure that you stay in focus for every point, that's why autofocus solutions have been developed by the manufacturers, using different principles.

Autofocus before the Raman image acquisition:

To calculate or measure the topography of the whole sample and then follow it during the measurement, by:

- Optical means
- Image treatment (like looking at video sharpness)
- If it's a plane (or a known surface), focus on three points (manually or using the autofocus) will determine the inclination plane of the sample

Autofocus during the Raman image acquisition:

To adjust the focus at every point, before each individual acquisition, but this can be much longer for large images, by:

- Data treatment, by optimizing the signal (no hardware required except z-stage)
- Optical means

Note that the z information obtained by the autofocus can be used to generate a surfacic 3D image of the sample, which means a 2D Raman image plotted on a topography map.

2.10.2.2 3D mapping

If the sample is reasonably transparent, combining z mapping to x-y mapping will generate a volumetric 3D image, by stacking images acquired at different altitudes (Figure 2.35).

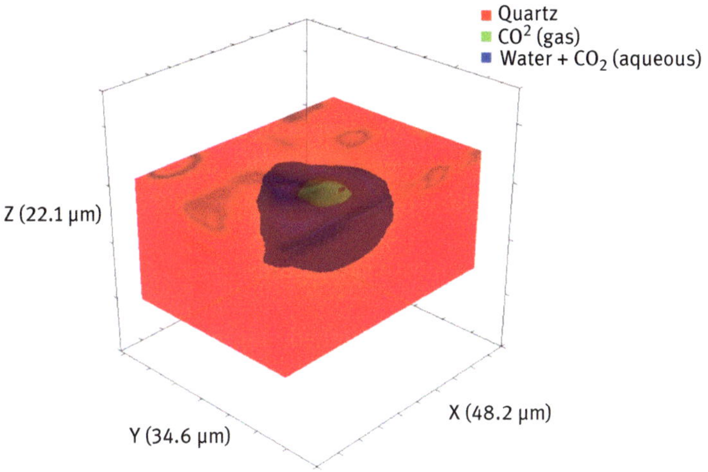

Figure 2.35: 3D volume map through fluid inclusion in a quartz matrix (step size 0.5 µm for X and Y, 1 µm for Z).

2.11 Other popular options

2.11.1 Polarization

For some applications it is important to analyze the response of the sample to different incident polarization states, for instance for crystal orientations etc. The Raman spectral bands can also be polarized and their analysis can reveal other properties.

To do so, there are two optical paths to consider:

- Laser path: usually the laser are polarized, let's say horizontally (H). To have a vertical polarization (V) it is necessary to insert on the laser path a component to convert H polarization into V, usually it is a halfwave plate (lambda/2 plate). If a circular polarization is desired, this is a quarter-waveplate to use.
- Raman (signal) path: to measure if a Raman band is polarized, a component called "analyzer" will be used on the Raman path, before the spectrometer. It's actually a regular polarizer, with its axis placed vertically or horizontally, depending which orientation has to be observed. But because the spectrometer has also a different response to H or V polarization, it is preferable to add a polarization scrambler (or depolarizer) between the polarizer and the spectrometer, in order to have equal responses to H or V polarized light.

2.11.2 UV Raman

For some applications where fluorescence is a big issue, UV Raman can be a solution. However, it requires specific hardware. NUV (325 nm and above) is reasonably accessible, whereas DUV (266 nm, 244 nm and below) is more difficult.

Special optics are required for UV coupling, because lenses designed for visible light will either not transmit UV light, or have strong focus shift, or both. Mirrors are achromatic by nature so in a multiwavelength system it's better to have only mirror optics, if you don't want to end up with a set of optics for each laser line (both on the Raman and the laser path). But even with mirrors, one must pay attention to its coating, which will generally be less efficient in DUV (unless you choose DUV dedicated mirrors).

The CCD detector has also to be chosen with care because usually the quantum efficiency drops down below 400 nm

The choice of microscope objectives is more limited, and lens-based objective will usually show strong longitudinal chromatic aberration (or focus shift), that will cause poor video image of the sample when the laser is in focus (and vice-versa). Thus, mirror based objectives are a good choice, but they must be carefully selected and aligned to provide good image.

Last, the lasers are more expensive, especially gas laser, which remains the main source of UV light (see Section 2.3).

In total, mirror efficiency, objectives efficiency, detector efficiency leads to significantly lower signal, therefore it is recommended to have enough laser power (usually above 50 mW at the laser output for DUV and 30 mW for NUV is recommended).

2.11.3 Coupling to SPM, TERS

Since several years it has become more and more easy to couple a SPM (Scanning Probe Microscope) to a Raman microscope and some manufacturers even propose combined systems.

The advantage of the SPM is its spatial resolution, which is typically the range of the nanometer, and can reach down to the atomic scale.

The first and easiest thing these combined systems can do is called "co-localized": it consists of doing a SPM image and overlaying it with a Raman image.

But the main interest is to achieve nanometric resolution Raman images, by using Tip Enhanced Raman Scattering (TERS). For this a more specific coupling system between both instruments is required, as well a specific SPM tips engineered for plasmonic effects which will enhance the Raman scattering only at the vicinity of the tip's apex, thus generating a very high spatial resolution Raman signal. Figure 2.36 is an example of such ultimate spatial resolution.

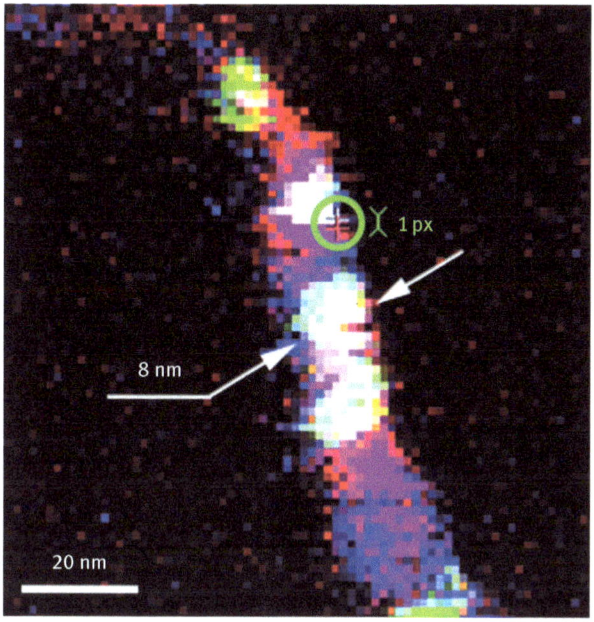

Figure 2.36: TERS image of a carbon nanotube showing local defects (white). The optical resolution is 8 nm, but the chemical sensitivity goes down to 1 pixel i.e., 1.3 nm.

2.11.4 Others

To close this chapter, you need to remember that the different Raman manufacturer are always improving their instruments and developing new options to increase the ease of use or the performance. Don't forget to visit their website or to ask for advice, as some options may not be shown, or could potentially be developed for your specific needs.

Some examples that have been recently seen:

- Reference sample
- Coupling to TCSPC or FLIM
- Optical tweezer
- Transmission Raman
- Photocurrent mapping
- External coupling to cryostats, magnetostats
- Coupling to SEM
- Electrochemistry
- Particle analysis
- Coherent Raman Scattering, like CARS or SRS

Izabella Jolan Jahn, Lydia Lehniger, Karina Weber,
Dana Cialla-May and Jürgen Popp

3 Sample preparation for Raman microspectroscopy

Abstract: Raman spectroscopy and its variants allow for the investigation of a wide range of biological and biomedical samples, i. e. tissue sections, single cells and small molecules. The obtained information is on a molecular level. By making use of databases and chemometrical approaches, the chemical composition of complex samples can also be defined. The measurement procedure is straight forward, however most often sample preparation protocols must be implemented. While pure samples, such as high purity powders or highly concentrated chemicals in aqueous solutions, can be directly measured without any prior sample purification step, samples of biological origin, such as tissue sections, pathogens in suspension or biofluids, food and beverages often require pre-processing steps prior to Raman measurements. In this book chapter, different strategies for handling and processing various sample matrices for a subsequent Raman microspectroscopic analysis were introduced illustrating the high potential of this promising technique for life science and medical applications. The presented methods range from standalone techniques, such as filtration, centrifugation or immunocapture to innovative platform approaches which will be exemplary addressed. Therefore, the reader will be introduced to methods that will simplify the complexity of the matrix in which the targeted molecular species are present allowing direct Raman measurements with bench top or portable setups.

Keywords: Raman spectroscopy, sample clean-up, pathogen, tissue, microfluidics, optical tweezers

3.1 Introduction

The Raman microspectroscopy enables the investigation of the entire range from tissue samples down to single cells to low molecular weight substances. It offers information on a molecular level and allows, based on the recorded spectrum, the identification of the chemical composition of the investigated samples. Pure samples, such as high purity powders or highly concentrated chemicals in aqueous solutions, can be directly measured without any prior sample purification step. Furthermore, in contrast to infrared spectroscopy, the presence of water does not

This article has previously been published in the journal *Physical Sciences Reviews*. Please cite as: Jahn, I. J., Lehniger, L., Weber, K., Cialla-May, D., Popp, J. Sample preparation for Raman Microspectroscopy *Physical Sciences Reviews* [Online] **2019**, 4. DOI: 10.1515/psr-2019-0018.

https://doi.org/10.1515/9783110515312-003

overshadow the Raman signal of the molecular species of interest. Thus, Raman spectroscopy can be easily performed on biological samples with high water content. This technique is also well known to be easily applicable for the identification of the composition of pharmaceuticals, geological and forensic samples.

Generally, the number of in-elastically scattered photons is approximately 1 out of 1 million incident photons and most of the samples need to be pre-concentrated in order to record high quality and reproducible Raman spectra of the targeted molecules. The acquired vibrational spectra present, depending on the spectral resolution, sharp and well separated Raman bands which can be unambiguously assigned to characteristic molecular vibrations. With increasing sample complexity, the recorded Raman spectrum will contain the spectral features of all Raman active molecules located in the focus volume. Therefore, the interpretation of the results becomes cumbersome and multivariate statistical analysis are required. The main alternative for this is sample clean-up in order to purify, isolate or enrich the targeted molecular species. Most of the gold standard methods i. e. high performance liquid chromatography or mass spectrometry applies extensive sample pre-processing steps, such as liquid–liquid extraction (LLE), protein precipitation or solid-phase extraction (SPE). Although, many of these procedures were also combined with Raman spectroscopy, the scientific community strives towards simplifying the clean-up procedures by taking advantage of the particular characteristics of this technique. Namely, no water interference, molecules containing chromophores will provide an enhanced Raman signal when the proper laser excitation wavelength is used or signals from single cells i. e. bacteria can be easily recorded.

Mostly samples of biological origin, such as tissue sections, pathogens in suspension or biofluids, food and beverages require sample preparation procedures prior to Raman measurements. Nonetheless, regardless of the sample origin, very often a substrate is used as sample holder. The material of these holders needs to be compatible with Raman measurements and has to be carefully selected by taking into consideration the measurement parameters [1]. For example, the wavelength of the employed laser excitation source will define the penetration depth of the photons and will restrict the use of conventional materials. Namely, the shorter the laser wavelength is the fewer signals will be collected from the supporting material. Therefore, for lasers emitting bellow ~ 550 nm conventional borosilicate glass slides can be used when the sample thickness is more than 5 µm. However, in most biological tissues auto fluorescence will be induced under these conditions, and a near infrared laser is most often opted for. In this case, monocrystalline CaF_2 glass or quartz is preferred as sample holder. Beside these, nickel foils, aluminum-covered silicon wafers or silicon were employed for Raman measurements to avoid background signals coming from the substrate.

In the present book chapter the most encountered sample preparation procedures that were combined with Raman microspectroscopy as detection method will be summarized. First, standalone techniques, such as filtration, centrifugation or immunocapture will be presented, whereas in the second part innovative platform approaches will be exemplary addressed. Therefore, the reader will be introduced to methods that will simplify the complexity of the matrix in which the targeted molecular species are present allowing direct Raman measurements with bench top or portable setups.

3.2 Sample clean-up and enrichment strategies

3.2.1 Filtration and centrifugation

One of the simplest methods for the separation of solid particles from a surrounding liquid is filtration. Depending on the object of investigation, a wide range of filters with different pore sizes can be applied. For instance, filtration is used for the separation of bacteria from eukaryotic cells due to their smaller size of 0.6–1.0 µm (for common bacteria). Moreover, it is also possible to perform several filtration steps one after another to achieve major purity. Stöckel et al. utilized two consecutive filtration steps to separate bacteria, more precisely *Burkholderia mallei* and *Burkholderia pseudomallei*, from animal fodder. First, suspensions were roughly filtrated through 20 threads gauze and finally with a 0.45 µm filter to remove remaining residues of the matrix. For subsequent Raman measurements, 1 µL of the suspension was dropped and air dried on nickel foil [2]. Filtration is a very common method for sample preparation, either as a sole sample preparation technique or as an important intermediate step for the further sample processing.

Another simple and effective technique for the separation of particles relies on centrifugation. There are two different types of centrifugation: differential centrifugation and density gradient centrifugation. The first mentioned method simply exploits the fact, that objects of different size or density will sediment at a diverse rate in a centrifuge.

The second technique, density gradient separation, is based on the sedimentation of particles in a density gradient and requires a special solvent. For instance, CsCl solution forms a stable density gradient after 24–48 h in a centrifuge at high speed. Dissolved macromolecules accumulate in one sheet, the so-called isopycnic zone, which corresponds to their buoyant density. This parameter can be measured after centrifugation in g/cm^3. CsCl solution covers a range from 1.3 to 1.8 g/cm^3, including most biomolecules like DNA (buoyant density: 1.7 g/cm^3), RNA (1.6 g/cm^3) or proteins (1.35–1.4 g/cm^3). Cell organelles which have a lower density can be separated by sucrose density gradients. Ficoll density gradient centrifugation is applied for the isolation of leukocytes from peripheral blood. Mononuclear cells can be collected from the resulting interphase between plasma and separation medium using a Pasteur pipet. Afterwards the cells are chemically fixed, suspended and their Raman spectrum can be recorded [3–7].

Density gradient centrifugation can also be utilized for the isolation of bacteria from complex matrices. Afterwards bacteria can be identified by single cell Raman measurements [8–10]. Another possible application is the separation of endospores from powder samples [11, 12]. Kloß et al. used centrifugation and subsequent Raman measurements to detect pathogens in a range of 10^3 cells/mL in patient urine samples [13].

Centrifugation and filtration steps can also be combined for an efficient sample preparation [14–16]. Collard et al. utilized an isolation method combining both techniques for the Raman spectroscopic detection of anthropogenic particles in fish stomach. Dissected stomach content was put into formaldehyde solution and filtered with a 5 µm cellulose acetate membrane filter. Afterwards the stomach content was digested overnight using NaClO solution. Another filter membrane was applied to collect the NaClO solution. Following another filtration step, the membrane was finally incubated in a methanol solution and exposed to ultrasounds. Particles were removed from the membrane and the solution could be centrifuged. Subsequently, the pellet was deposited on a stainless-steel plate and analyzed via Raman spectroscopy [16].

3.2.2 Immunocapture-like assays

Liquid samples of clinical or environmental relevance contain pathogens in suspension at a wide range of concentrations. When Raman spectroscopy is applied as measurement technique, due to its low sensitivity and low detection volume, the direct detection in the raw sample of these pathogens will be challenging. One way to improve the sensitivity and speed of Raman-based platforms is to capture these organelles on surfaces functionalized with antibodies. For instance, antigen–antibody like interactions are known to be high affinity key-lock type bindings and their combination with Raman and surface enhanced Raman spectroscopy (SERS) proved to offer great advantages over the spectroscopy methods alone [17–19].

In a classical immunosensor, a given antibody is selected to specifically capture one antigen. Nevertheless, this can turn into a disadvantage if multiple targets are needed to be detected from the same sample. The development of specific antibodies for each target is time consuming and cost intensive. Furthermore, cross reactivity is also a very well-known disadvantage of immunosensors. By employing Raman for the reliable identification of the captured antigens, antibodies against common cell wall structures in Gram-positive and Gram-negative bacteria can be used instead of very specific antibodies [17]. In this way, a wide variety of pathogens can be isolated at once. In order to achieve this, a chip with aluminum fields was functionalized with an organosilane in order to provide reactive epoxy groups for the covalent attachment of antibodies through their amine, thiol or hydroxyl moieties. Aluminum is the support material of choice as it will not interfere with the Raman signal of the bacteria, while a covalent bond was selected in order to endure multiple washing steps to remove molecules un-specifically bound to the surface. The Raman measurements could be performed after 10 min incubation of the chip in the sample containing different strains and types of bacteria, at concentrations between 5×10^8 and 5×10^3 cells/mL. An alternative assay design allows for fast, simple and label-free Raman detection of bacteria from only a sample droplet of a few microliters by employing planar microelectrode arrays (Figure 3.1(A)) [18]. The alternating current electric field generated by the array causes an accelerated bacterial transport from the droplet to the surface modified with antibodies. After 15 min at 10 kHz the bacteria were collected and the chips were rinsed with a buffer and water and then subjected to Raman measurements. Here, silica was used as support material for the platform; this being another material giving rise to only a few well-defined Raman bands, and thus it will not interfere with the Raman signal of the isolated bacteria.

Bacteria when successfully isolated can be easily identified based on their Raman signature. Nevertheless, when it comes to biomarkers present in low concentrations in the samples, the number of inelastically scattered photons is too low to deliver a reliable signal. One way to enhance the weak Raman signal is by applying plasmonic nanoparticles. The number of SERS-based immunoassays significantly increased during the last decade. Generally, biomarkers are isolated from samples by conventional immunocapture strategies while their presence is confirmed by the Raman spectra of the so-called SERS-tags or SERS-immunoprobes [19]. The available designs range from homogenous to heterogeneous, from non-competitive to competitive ones and labeled and label free ones. Among the targeted molecules proteins, nucleic acids, viruses, cells and toxins were in the focus. For example, a competitive SERS-based immunoassay allows the simultaneous detection of cTNI and CK-MB, two cardiac biomarkers in human blood, within 15 min and with a 95 % limit of agreement with a commercial assay [20]. For this, the two monoclonal antibodies were conjugates with magnetic beads, while the target antigens were immobilized on the surface of the SERS tags (Figure 3.1(B)). In this way, the antigen conjugated SERS tags will compete with the free biomarkers from the sample for binding sites on the magnetic surface. After only 7 min interaction time a magnetic bar was used to separate the magnetic particles from the solution and the remaining supernatant was measured with SERS. The limit of detection was 42.5 and 33.7 pg/mL for CK-MB and cTnI, respectively.

Overall, antibody–antigen based reactions offer high specificity and Raman spectroscopy can easily confirm the presence of the captured targets. Because of the strong existing bonds, the non-specifically bound molecules can be successfully washed off and they will not interfere with the Raman signal of the targets. Nevertheless, bacteria are constantly mutating and adapting to their environment, hence, as a consequence it may happen that after mutation they cannot be anymore captured by the already available antibodies. A solution for this challenge is

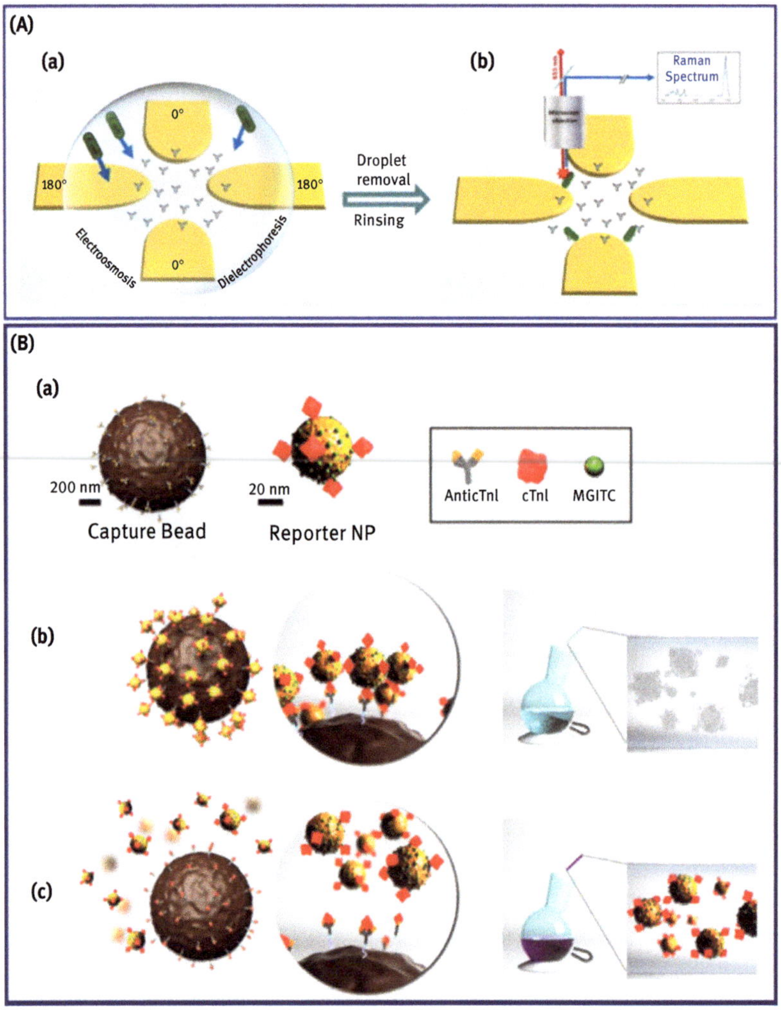

Figure 3.1: (A) Bacteria capturing and identification: (a) Bacteria suspended in a droplet are driven to accumulate at the tips of an energized quadrupole microelectrode array. (b) Following droplet removal and rinsing with water, micro-Raman spectroscopy is used to probe each of the four inter-electrode gaps along the line of minimum separation. Reproduced from Ref [18]. (B) Schematic illustration of the SERS-based competitive immunoassay for quantification of cardiac biomarkers: (a) Preparation of antibody-conjugated capture magnetic beads and SERS tags. Interactions between SERS tags and capture magnetic beads under (b) low and (c) high concentrations of free cTnI antigen in solution [Reproduced Ref [20] with permission of The Royal Society of Chemistry].

to use recognition elements less susceptible to mutation, such as glycanes and siderophores. These surface structures are directly related to the pathogens' virulence and are referred to as "immutable ligands". Therefore, the same chip presented above with aluminum fields was in a subsequent study also functionalized with pyoverdine, an iron rich substance excreted by fluorescent pseudomonas [21]. *Pseudomonas aeruginosa* and *Pseudomonas fluorescence* were successfully captured by this platform from tap water samples and identified based on their Raman signature.

3.2.3 Liquid–Liquid Extraction (LLE)

LLE is a conventional sample separation method, which exploits the different solubility of substances in two immiscible solvents. The principle is based on transferring the solvate (or solvates) present in the primary liquid solution to another immiscible liquid (solvent). The solvate-enriched solvent is referred to as *extract*, the diluted starting solution as *raffinate*. Often water is used as hydrophilic solvent as well as a second hydrophobic, organic solvent.

The solution containing the component of interest is mixed with the solvent in a separating funnel by vigorous shaking. This leads to an increase in the phase boundary between the two solvents. According to the Nernst's distribution law, a distribution between the two phases occurs, which remains constant when in equilibrium. Depending on the equilibrium constant, a certain amount of substance will pass into the extractant. At rest, two layers are formed again according to the different densities of the solvents. After separation, the product can be recovered by evaporation of the solvent. By repeated addition of the extractant and re-separation, the desired product can be almost completely isolated and subsequently easily measured via Raman. For instance, Thien et al. developed a measurement setup combining the advantages of microfluidics and Raman microspectroscopy for a time and material efficient determination of LLE data [22].

3.2.4 Solid-phase extraction (SPE)

SPE represents another well-known sample preparation technique, which enables the enrichment, concentration and isolation of analytes through specific interaction on a solid phase (sorbent). Usually the functionalized sorbent is fixed in a column or cartridge. The sample and a suitable solvent pass the column with the help of pressure or a vacuum. SPE encompasses the following steps: conditioning of the sorbent, sample application, washing to remove interfering components, drying and elution of the analyte. SPE can rely upon many different ways of molecule interactions: polar/nonpolar, ionic, covalent binding, etc. Choosing the optimal sorbent depends crucially on the sample solution. In case of aqueous sample solutions, only nonpolar sorbents can be used, otherwise the water molecules would occupy the available adsorption positions on the surface of the sorbent. To extract analytes from nonpolar solutions, polar sorbents like silica gel are preferable.

After removing the adsorbed analyte from the solvent phase, quantification is realized by gas chromatography, mass spectrometry, high-performance liquid chromatography or Raman spectroscopy.

Raman spectroscopy often demands the pre-concentration of small volumes prior to detection and quantification. One major drawback of SPE is having to resolve or wash the analyte into another solution volume for the detection. Kitt et al. avoided this step by using a single chromatographic particle as an extractor [23]. Confocal Raman microscopy was applied for in situ detection within the single particle collection phase. Specifically, micromolar concentrations of pyrene in methanol/water solution were equilibrated with an individual C18-functionalized silica particle, and Raman spectra were acquired from a small confocal sampling volume (~1 fL) within the particle interior [23].

Nwaneshiudu et al. combined solid-phase micro-extraction (SPME) with Raman spectroscopy (SPME-RS) to detect small target analytes. Polydimethylsiloxane (PDMS)-based solid-phase micro-extraction was used along with Raman spectroscopy to separate and enhance the detection of five anesthetic compounds (halothane, propofol, isoflurane, enflurane and etomidate) from aqueous and serum phases [24].

Owing to its low cost, ease of use, and nonpolluting means of preparing samples for analysis, SPE is fast replacing traditional liquid-liquid methods in clinical, pharmaceutical, agricultural and industrial applications.

3.2.5 Tissue processing

Raman spectroscopy can extract high quality chemical information from biological samples and can be used for imaging of a large variety of tissue sections for medical diagnosis purposes. Most of the published approaches handle *ex-vivo* tissues and only a few are directed towards *in-vivo* applications. Vibrational spectroscopy is very sensitive to structural changes of the tissue constituting molecules; hence, it is very important to maintain tissue integrity during sample preparation. This is also vital for the development of spectral histopathology because a spectral database obtained on tissues needs to be first established before disease determination can be reliably done on fresh tissue.

Generally, two preservation methods are applied for surgically excised tissue: (1) paraffin embedding or (2) snap-freezing [25]. The first one, formalin fixation and paraffin preservation, is more prevalent compared to the latter one, although, the as-prepared tissues are not always compatible with Raman microspectroscopy. Namely, the tissue is initially immersed in an aqueous formalin solution. Formalin cross-links the primary and secondary amine groups of proteins [26], whereas by reacting with the double bonds of unsaturated hydrocarbon chains it preserves some of the lipids [27]. Afterwards, by immersing the tissue in solutions with increasing ethanol concentrations the sample is dehydrated. This can denaturate the tertiary structure of proteins and can lead to the coagulation of globular proteins and lipids that were not preserved via formalin can precipitate. After dehydration, the sample is immersed in xylene and subsequently transferred to molten paraffin wax. The wax penetrates through the tissue due to the existing pores created in the dehydration process. The as-prepared samples can be stored for a long time in biobanks and can be sectioned (2–10 µm) with a microtone prior measurement. Because paraffin penetrates through the tissue, the Raman spectra recorded on these samples will present bands that can be assigned to the vibrational modes of the molecules constituting the wax. This can overshadow the tissue characteristic Raman bands and will reduce the diagnosis value of the method. Additionally, by taking into account the structural changes of proteins and lipids mentioned above, it can be concluded, that the obtained spectral information will differ from the one of the native tissue. While dewaxing procedures could reduce the interfering signal of the wax matrix [28–30], the latter effect cannot be avoided.

Snap-freezing is the second most commonly used method for preserving tissues. When no embedding matrix, such as an optimal cutting temperature compound is used, the Raman bands observed in the measured spectra can be unambiguously assigned to tissue components. This is a great advantage over the previously described preservation method. Nevertheless, there are multiple studies trying to assess whether snap-freezing and multiple freeze-thaw cycles can affect the tissue in such a manner that this could lead to significant spectral differences [31–33]. Namely, it is known that the water contained in the tissue acts as supporting medium. When water is slowly frozen it can exist in two different crystalline forms. However, when the cooling process is instantaneous, a vitreous matrix can be also obtained [25]. The latter is preferred, as during this process the water does not have enough time to expand, and thus, cell damage is minimized. However, care should be taken, as for large tissue blocks the vitreous phase will form only at the surface, while the inner regions of the sample will be subjected to lower freezing rate, hence, to crystal formation. Because vitreous ice is an unstable phase, tissue sectioning should be carried out as soon as possible after snap-freezing. Generally, before cryosectioning is performed, the samples have to be warmed up above the snap-freezing temperature in order to avoid the tissue to shatter. Therefore, water expansion cannot be fully avoided. In a published study, where neuroblastoma and other pediatric neural crest tumors were analyzed with Raman spectroscopy, the authors show that cryopreserved tissue can be accurately classified [31]. According to their results, the Raman spectra recorded on fresh and on frozen tissue show high correlation. Therefore, spectral databases build on snap-frozen tissues from biobanks could be used for medical diagnosis of fresh neuroblastoma. In another study, where different fixation methods for parenchymal tissue from placenta were investigated [34], the authors noticed several spectral differences in the Raman modes ascribed to proteins. This might be caused by the depolymerization of the cellular skeleton, and consequently by the protein structural changes. Overall, care must be taken when spectral databases on frozen tissues are established and further used for medical diagnosis.

3.3 Innovative platform approaches

3.3.1 Microfluidics

As discussed in the introduction of this chapter, pure liquid samples rarely require clean-up in order to identify their chemical composition based on the measured Raman spectrum. Nevertheless, when it comes to bodyfluids such as blood or urine, collected from patients, most often the detection of a biomarker is aimed for and the presence of the other constituents of the sample might inhibit this process. Therefore, extensive sample preparation steps are undertaken; most of them were described in the previous section. In clinical laboratories, reduced sample volume requirement and automation are, among others, two of the most common requirements for analytical platforms. Ideally, in order to increase patient comfort, a drop of blood should be enough in order to diagnose or monitor diseases. In the last decades, the transition from large sample volumes to just micro- or nanoliter amounts was facilitated by the development and implementation of micro/nanofluidic systems into the entire analysis chain. By employing these platforms both sample volume and waste amounts are considerably reduced and at the same time the automation is significantly increased. Up to date, microfluidic platforms come in various sizes and forms and incorporate multiple active or passive functional units. They mainly serve as sample handling units and are connected with various optical and electrical read-out techniques.

In clinical microbiology, depending on the pathogen, the turnaround time of the assays range between 24 h to weeks due to the required culturing steps. Additionally, in the case of non-cultivable pathogens new challenges appear. Within Raman microspectroscopic techniques bacteria can be identified at single cell level, thus it can be carried out on the raw sample. Nevertheless, the low concentration of the bacteria in suspension makes the direct Raman detection challenging. Multiple ways to capture and enrich bacteria directly from bodyfluids using microfluidic platforms have been reported and some of the most promising approaches that were combined with Raman detection will be presented in the following.

Microfluidic platforms are often combined with dielectrophoresis in order to capture bacteria from liquid samples. This approach is also suitable for Raman detection and it allows an automated control of pathogens with increased need for biosafety. For instance, by using a quadruple electrode design with negative dielectrophoretic trapping, uropathogens were successfully trapped and concentrated from urine samples for further Raman spectroscopic analysis [35]. The method relies on the polarization effects of neutral matter in non-uniform electric fields. For the successful capturing, the applied voltage on the implemented electrodes has to be carefully adjusted by taking into account the conductivity of the urine samples. In this way, after the urine sample is collected from patients and an off chip filtration step is carried out to remove cellular debris a low volume of sample, ~ 200 µL, can be injected into the microfluidic platform for bacteria trapping and Raman analysis. An alternative approach for pathogen trapping from urine samples can be achieved with the "Lab-on-a-Disc" platform depicted in Figure 3.2(A) [36]. A glass-polymer hybrid chip mounted on a rotor facilitates the concentration of bacteria from suspension in V-shaped structures due to the induced centrifugal forces. In this way, after stopping the rotation, Raman spectra can be acquired in a couple of seconds from the now stationary bacteria. The limit of detection was above 2×10^7 cells/mL within a 4 µL sample volume. This platform allows the processing of four individual samples, while still being able to trace the position of each sample. The holder was designed in such a way that the loading, adjustment and unloading of the V-shaped chips can be achieved in a couple of seconds. Compared with pressure driven pumps, centrifugal microfluidic platforms are more robust as the separation scales only with the sample density and the angular frequency of the rotation, while compared with the previous approach the electrical conductivity of the sample does not play a role anymore. However, off-chip processing steps such as filtration in order to get rid of large particles present in the urine samples, volume reduction and

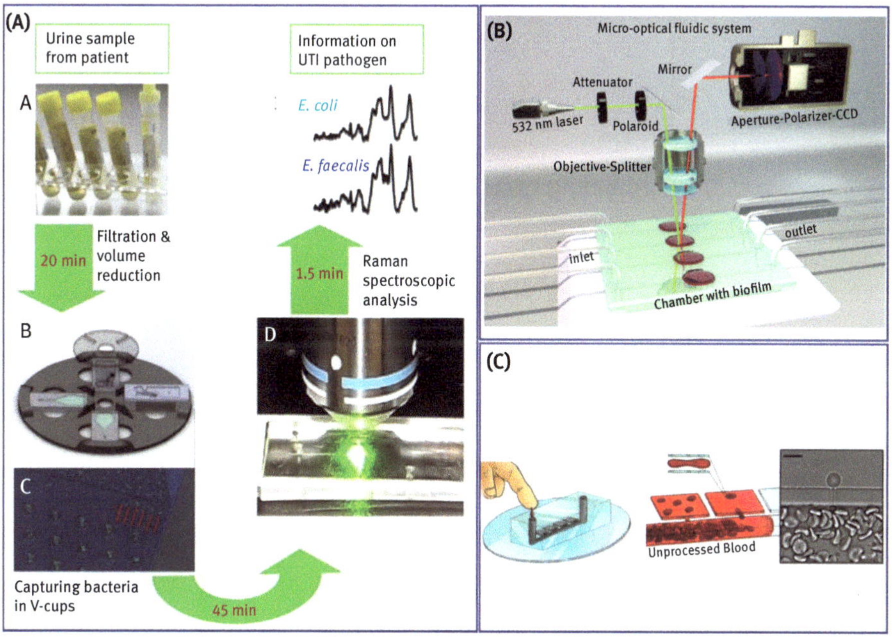

Figure 3.2: (A) "Lab-on-a-Disk" for uropathogen identification from suspension [Reprinted from Ref [36] with the permission of AIP Publishing. Copyright © 2015 AIP Publishing LLC.]; (B) Microfluidic array for continuous characterization of *Pseudomonas aeruginosa* biofilms [Reproduced from Ref [37] with permission of The Royal Society of Chemistry]; (C) Self-filling microfluidic device for red blood cell investigations Reprinted from Ref [38], with the permission of AIP Publishing].

medium exchange to avoid fluorescence interference with the Raman signal had to be applied prior to the analysis for both approaches. The later step could have been omitted when e. g. a 785 nm excitation laser instead of the 532 nm would be used in order not to excite fluorescence emission.

In the previous examples cells were captured from suspension and analyzed in bulk. However, it is of high interest to be able to sort the cells and collect them separately, while keeping their integrity for further analysis. Raman-activated cell sorting is a newly emerging technique as it allows, in contrast to fluorescence, a label free identification and sorting of particles. This can be achieved by either applying dielectrophoresis for capturing [39] or pressure dividers for continuous flow sorting [40]. In the first case, positive dielectrophoresis geometry was combined with solenoid-valve-suction-based switch for single cell trapping, releasing and automatic separation at sub-second level at high flowrates. By applying a periodical field gradient at the electrodes, yeast cells were trapped, ordered and delivered edge-to-edge to the laser detection point. The trapping duration could be adjusted between 90 ms to minutes in order to allow the acquisition of high quality Raman spectra and to sort yeast cells from a mixture. In the second approach, pressure dividers and hydrodynamic flow focusing were used for delivering the cells to the site of Raman detection. Overall, single cell manipulation is very attractive, nevertheless the robustness of these techniques still has to be tested for different pathogen mixtures and their limit has to be pushed toward clinical samples, where a high diversity of particles will be present and might inhibit the Raman detection of the targeted pathogens.

Besides single cell analysis, microfluidic platforms offer the possibility to monitor the formation and metabolism of biofilms by Raman detection in a controlled and non-destructive manner [37, 41]. This combination offers the possibility to continuously provide nutrient broth and to investigate the chemical composition of biofilms over time automatically (Figure 3.2(B)). Furthermore, by employing a microfluidic array with parallel culture chambers, the response of cells to a large number of different stimuli can be monitored on one platform for different replicates. In this way i. e. the quantitative investigation of lipid droplet morphology in live cells *in-situ* under 10 different chemical environments and with 8 replicates for each culture condition could be achieved [41] by applying stimulated Raman spectroscopy as detection technique.

Lastly, single red blood cell experiments, where Raman spectroscopy is used to interrogate the chemical information, can be also performed employing unprocessed whole blood obtained by finger pricking by using a microfluidic assay (Figure 3.2(C)) [38]. A self-filling chip with specialized microfluidic geometry allows exploiting the unique mechanical properties of red blood cells in order to autonomously separate and isolate them from the bulk into diffusion controlled micro-chambers. Due to the low height of these micro-chambers, the cells are confined and they can not move out of the focus plane of the Raman spectrometer. After the self-filling process is finished, a washing step with buffers regulated by advection can be performed. Therefore, the system can be used to study the behavior of red blood cells to external stimuli without compromising their integrity by trapping with electrical or optical fields.

Overall, pathogen trapping and sorting, monitoring of biofilm formation and cell metabolism or the study of the chemical changes induced by external stimuli on red blood cells can be automatically performed with minimal sample volume requirement by applying microfluidic platforms for sample clean-up and handling. Extensive work still needs to be done in order to push the limits of these techniques toward high-throughput measurements on clinical samples; nevertheless, the recent publications demonstrate their high potential.

3.3.2 Optical tweezers as tool in Raman microspectroscopy

In order to trap dielectric particles, gradient forces are applied i. e. the particle will move to the maximum field intensity in the case that it possesses a higher dielectric constant than the medium [42]. The optical tweezer technique is based on this concept and is nowadays utilized in biology, biophysics and medicine in order to manipulate microparticles by using light. The physical background is summarized in the literature to which the interested reader is referred to [42, 43]. By the combination with Raman spectroscopy, molecular specific fingerprint information is recorded of the trapped specimens and thus, the optical tweezer act as a sample preparation step in order to enrich or to manipulate and capture the analyte of interest within the laser focus.

The estimation of the polymer concentration in polymer-rich domains is illustrated by using the example of phase-separated poly-(N-isopropylacrylamide) aqueous solutions [44]. Here, the optical tweezer traps the polymer-rich-domain within the focus of the laser light and the Raman spectra are recorded. In the case of heterogeneous catalysis and atmospheric chemistry, the observation of the interaction of gas molecules with solid microparticles over a long timescale is of high importance. To achieve experiment periods up to 24 h, the optical trap is optimized by using a counter propagating dual beam for the levitation of solid particles in air [45]. In this study, the adsorption of water molecules onto silica particles is investigated by increasing the humidity within the trapping chamber. Thus, the optical tweezer is well suited for trapping the specimens of interest in order to perform Raman investigations on single microparticle level.

The beneficial combination of the optical tweezer's concept with Raman microspectroscopy is summarized for the investigation of biological microparticles, such as cells, spores and aerosols, in [43]. As concluded by the authors of the review article, the Raman spectroscopic investigation of trapped microparticles allows for the detection on single particle level and the identification of the influence of environmental changes onto the individual trapped biological specimen. To administer pharmaceutical drugs for treating lung diseases, particle inhalation is an effective tool [46]. Since the humidity is changed from the inhaler to the lung, hygroscopic phase transitions or particle growth might occur. Thus, the uptake of the bronchodilation drugs could be negatively affected resulting in an inefficient therapy. In order to characterize the hygroscopic properties of pharmaceutical drugs, the aerosol particles were generated by commercial available inhalers and captured in an optical tweezer allowing for online Raman measurements. In the case of salmeterol xinafoate only small and for fluticasone propionate as well as ciclesonide no structural changes were observed with increasing humidity. The arterial blood pH value is monitored by trapping single live erythrocytes and recording their Raman spectra [47]. The obtained in vitro results show, that the Raman intensity ratio of the C=C in plane bending mode and the C=C stretching mode decreases within the pH value range from 6.5 to 9. This study shows the potential of laser tweezer Raman spectroscopy in medical diagnosis since the Raman signal of an individual erythrocyte cell is a biomarker of the extracellular pH value. Moreover, the Raman fingerprint information of red blood cells is applied for ABO blood typing [48]. The Raman spectra of the trapped cells are analyzed by using principal component analysis in combination with linear discriminant analysis to identify the blood type. Furthermore, Raman optical tweezers are applied for the investigation of cells in different phases of their cell cycle [49]. Since the densest part of the cell will be in the laser focal point, the application of this technique results in the domination of the Raman features of the cell nucleus without prior extraction or isolation. Finally, the huge potential of the laser trapping Raman spectroscopic approach is illustrated by the investigation of extracellular vesicles such as exosomes, nanoscale particles excreted by cells and with important properties in cell signaling [50]. They differ in a large extent in composition, size and most likely also in biological function, which illustrates the need for a molecular specific detection scheme on single particle level. Here, a multispectral optical tweezer is utilized allowing for both Raman as well as fluorescence spectroscopic detection. The authors observed that nucleic acids, proteins, lipids and cholesterol dominate the Raman spectra of exosomes. By functionalizing the CD9+ subpopulation with a fluorescent dye, these vesicles were selectively trapped and their Raman spectrum is recorded. In comparison to the bulk vesicle population as isolated by classical ultracentrifugation, the Raman fingerprint signals provide information that the CD9+ exosome subset exhibited less chemical heterogeneity and also reduced component concentration.

In order to improve the trapping properties for nanoscale specimens such as single protein molecules, nanoaperture optical tweezers were developed, which allow for a 1000 times more efficient trapping than conventional optical tweezers [51]. The theoretical background is well summarized within this review article illustrating the various geometries of nanoaperture traps such as circular, rectangular, double nanohole and bowtie arrangements. The same group published a communication on the Raman spectroscopic characterization of single titania and polystyrene nanoparticles employing a double-nanohole optical tweezer approach [52]. The identification of the trapping event is achieved by recording the fingerprint Raman spectra of individual nanoparticles showing the specific vibrational modes of the components when comparing with bulk measurements. Finally, the trapping properties of optical tweezers are improved by standing-wave Raman tweezers for the sensitive and specific characterization of isolated nanostructures applying less laser intensity compared to conventional trapping experiments [53]. The authors demonstrated the potential of their proposed technique by analyzing individual single-wall carbon nanotubes, graphene flakes, biological particles or SERS-active metal nanoparticles.

To conclude, the combination of optical tweezers with Raman spectroscopy offers a powerful tool to capture dense structures within the focal point and thus allowing for the analysis without prior isolation of the specimen of interest. By employing emerging techniques such as nanoaperture optical tweezers, the detection and characterization of individual nanostructures such as protein molecules are achieved.

3.4 Future perspectives and conclusion

In this chapter, different strategies for handling and processing various sample matrices for a subsequent Raman microspectroscopic analysis were introduced illustrating the high potential of this promising technique for life science and medical applications.

Raman spectroscopy and its variants allow for the investigation of the entire range of biological and biomedical samples, i. e. from tissue samples to single cells and small molecules. Statistical models and databases are often used for the identification of the sample, e. g. bacterial cells, tumor margins or low-molecular weight substances. The most time-consuming step in the entire analysis chain is often sample preparation. For example, in order to perform Raman measurements on cells, the cells usually first have to be isolated from their surrounding matrix. Depending on the sample composition, this task can be very demanding. Moreover, the direct investigation of trace level concentrations of the target molecule via Raman is challenging and needs specific pre-processing steps.

Nevertheless, numerous approaches have been developed in recent years that allow the investigation of target analytes in different matrices using Raman microspectroscopy. In addition to mechanical strategies such as filtration and centrifugation, microfluidic devices, chip-based systems or optical tweezers have proven to be Raman-compatible. In order to promote the application of Raman-based detection systems in life science, the implementation of standard methods would be a useful step to map the entire process chain from sampling and isolation to Raman measurement parameters and chemometric identification. However, the variety of different isolation and identification strategies presented in the chapter shows, that at this stage of knowledge, an individual solution has to be found for each specific bioanalytical question. Extensive work still needs to be done in order to push the limits of these techniques toward high-throughput measurements for instance on clinical samples; nevertheless, the recent publications demonstrate their high potential and the broad applicability in life sciences.

References

[1] Ramoji A, Galler K, Glaser U, Henkel T, Mayer G, Dellith J, et al. Characterization of different substrates for Raman spectroscopic imaging of eukaryotic cells. J Raman Spectrosc. 2016;47:773–86.

[2] Stöckel S, Meisel S, Elschner M, Melzer F, Rösch P, Popp J. Raman spectroscopic detection and identification of Burkholderia mallei and Burkholderia pseudomallei in feedstuff. Anal Bioanal Chem. 2015;407:787–94.

[3] Ramoji A, Neugebauer U, Bocklitz T, Foerster M, Kiehntopf M, Bauer M, et al. Toward a spectroscopic hemogram: Raman spectroscopic differentiation of the two most abundant leukocytes from peripheral blood. Anal Chem. 2012;84:5335–42.

[4] Workalemahu G, Foerster M, Kroegel C. Expression of metalloproteinase-7 (matrilysin) in human blood and bronchoalveolar gamma/delta T-lymphocytes. Selective upregulation by the soluble non-peptidic mycobacterial phosphoantigen (isopentenyl pyrophosphate). J Cell Physiol. 2006;207:67–74.

[5] Neugebauer U, Bocklitz T, Clement JH, Krafft C, Popp J. Towards detection and identification of circulating tumour cells using Raman spectroscopy. Analyst. 2010;135:3178–82.

[6] Pachmann K, Clement JH, Schneider CP, Willen B, Camara O, Pachmann U, et al. Standardized quantification of circulating peripheral tumor cells from lung and breast cancer. Clin Chem Lab Med. 2005;431434-6621 (Print)):617–27.

[7] Schwalbe M, Pachmann K, Höffken K, Clement JH. Improvement of the separation of tumour cells from peripheral blood cells using magnetic nanoparticles. J Phys Condens Matter. 2006;18:S2865.

[8] Pahlow S, Meisel S, Cialla-May D, Weber K, Rösch P, Popp J. Isolation and identification of bacteria by means of Raman spectroscopy. Adv Drug Deliv Rev. 2015;89:105–20.

[9] Meisel S, Stöckel S, Elschner M, Rösch P, Popp J. Assessment of two isolation techniques for bacteria in milk towards their compatibility with Raman spectroscopy. Analyst. 2011;136:4997–5005.

[10] Meisel S, Stöckel S, Elschner M, Melzer F, Rösch P, Popp J. Raman spectroscopy as a potential tool for detection of brucella spp. in milk. Appl Environ Microbiol. 2012;78:5575–83.

[11] Stöckel S, Meisel S, Elschner M, Rösch P, Popp J. Identification of bacillus anthracis via Raman spectroscopy and chemometric approaches. Anal Chem. 2012;84:9873–80.

[12] Stöckel S, Meisel S, Elschner M, Rösch P, Popp J. Raman spectroscopic detection of anthrax endospores in powder samples. Angew Chem Int Ed. 2012;51:5339–42.

[13] Kloss S, Kampe B, Sachse S, Rösch P, Straube E, Pfister W, et al. Culture independent Raman spectroscopic identification of urinary tract infection pathogens: A proof of principle study. Anal Chem. 2013;85:9610–16.

[14] Meisel S, Stöckel S, Rösch P, Popp J. Identification of meat-associated pathogens via Raman microspectroscopy. Food Microbiol. 2014;38:36–43.

[15] van de Vossenberg J, Tervahauta H, Maquelin K. Identification of bacteria in drinking water with Raman spectroscopy. Analytical Methods. 2013;5:2679–87.
[16] Collard F, Gilbert B, Eppe G, Parmentier E, Das K. Detection of anthropogenic particles in fish stomachs: An isolation method adapted to identification by Raman spectroscopy. Arch Environ Contam Toxicol. 2015;69:331–9.
[17] Pahlow S, Kloß S, Blaettel V, Kirsch K, Hübner U, Cialla D, et al. Isolation and enrichment of pathogens with a surface-modified aluminium chip for raman spectroscopic applications. ChemPhysChem. 2013;14:3600–5.
[18] Liao DS, Raveendran J, Golchi S, Docoslis A. Fast and sensitive detection of bacteria from a water droplet by means of electric field effects and micro-Raman spectroscopy. Sens Bio-Sens Res. 2015;6:59–66.
[19] Wang Z, Zong S, Wu L, Zhu D, Cui Y. SERS-activated platforms for immunoassay: Probes, encoding methods, and applications. Chem Rev. 2017;117:7910–63.
[20] Chon H, Lee S, Yoon S-Y, Lee EK , Chang S-I, Choo J. SERS-based competitive immunoassay of troponin I and CK-MB markers for early diagnosis of acute myocardial infarction. Chem Commun. 2014;50:1058–60.
[21] Pahlow S, Stöckel S, Pollok S. Rapid identification of pseudomonas spp. Via Raman spectroscopy using pyoverdine as capture probe. Anal Chem. 2016;88:1570–7.
[22] Thien J, Peters C, Brands T, Koß H-J, Bardow A. Efficient determination of liquid–liquid equilibria using microfluidics and Raman microspectroscopy. Ind Eng Chem Res. 2017;56:13905–10.
[23] Kitt JP, Harris JM. Confocal Raman microscopy for in situ detection of solid-phase extraction of pyrene into single c-18-silica particles. Anal Chem. 2014;86:1719–25.
[24] Nwaneshiudu IC, Nwaneshiudu CA, Schwartz DT. Separation and enhanced detection of anesthetic compounds using solid phase micro-extraction (spme)-Raman spectroscopy. Appl Spectrosc. 2014;68:1254–9.
[25] Lyng F, Gazi E, Gardner P. Chapter 5 Preparation of tissues and cells for infrared and Raman spectroscopy and imaging, In Biomedical applications of synchrotron infrared microspectroscopy: A practical approach. In: David Moss, editor. The Royal Society of Chemistry, 2011:145–91. Cambridge, UK:
[26] Thavarajah R, Mudimbaimannar VK, Elizabeth J, Rao UK, Ranganathan K. Chemical and physical basics of routine formaldehyde fixation. J Oral Maxillofac Pathol: JOMFP. 2012;16:400–5.
[27] Jones D. Reactions of aldehydes with unsaturated fatty acids during histological fixation. In Stoward PJ, editor. Fixation in Histochemistry. Boston, MA: Springer US, 1973:1–45.
[28] Faoláin EÓ, Hunter MB, Byrne JM, Kelehan P, Lambkin HA, Byrne HJ, et al. Raman spectroscopic evaluation of efficacy of current paraffin wax section dewaxing agents. J Histochem Cytochem. 2005;53:121–9.
[29] Huang Z, McWILLIAMS AN, Lam S, English J, McLEAN DI, Lui H, et al. Effect of formalin fixation on the near-infrared Raman spectroscopy of normal and cancerous human bronchial tissues. Int J Oncol. 2003;23:649–55.
[30] Fullwood LM, Griffiths D, Ashton K, Dawson T, Lea RW, Davis C, et al. Effect of substrate choice and tissue type on tissue preparation for spectral histopathology by Raman microspectroscopy. Analyst. 2014;139:446–54.
[31] Wills H, Kast R, Stewart C, Rabah R, Pandya A, Poulik Jk, et al. Raman spectroscopy detects and distinguishes neuroblastoma and related tissues in fresh and (banked) frozen specimens. J Pediatr Surg. 2009;44:386–91.
[32] Shim MG, Wilson BC. The effects of ex vivo handling procedures on the near infrared Raman spectra of normal mammalian tissues. Photochem Photobiol. 1996;63(0031-8655 (Print)):662–71.

[33] McElderry JD, Kole MR, Morris MD. Repeated freeze-thawing of bone tissue affects Raman bone quality measurements. J Biomed Opt. 2011;16:071407.
[34] Faolain EO, Hunter MB, Byrne JM, Kelehan P, McNamara M, Byrne HJ, et al. A study examining the effects of tissue processing on human tissue sections using vibrational spectroscopy. Vib Spectrosc. 2005;38:121–7.
[35] Schröder UC, Ramoji A, Glaser U, Sachse S, Leiterer C, Csaki A, et al. Combined dielectrophoresis–Raman setup for the classification of pathogens recovered from the urinary tract. Anal Chem. 2013;85:10717–24.
[36] Schröder UC, Bokeloh F, O'Sullivan M, Glaser U, Wolf K, Pfister W, et al. Rapid, culture-independent, optical diagnostics of centrifugally captured bacteria from urine samples. Biomicrofluidics. 2015;9:12.
[37] Feng J, de la Fuente-Nunez C, Trimble MJ, Xu J, Hancock REW, Lu X. An in situ Raman spectroscopy-based microfluidic "lab-on-a-chip" platform for non-destructive and continuous characterization of pseudomonas aeruginosa biofilms. Chem Commun. 2015;51:8966–9.
[38] Gollner M, Toma AC, Strelnikova N, Deshpande S, Pfohl T. A self-filling microfluidic device for noninvasive and time-resolved single red blood cell experiments. Biomicrofluidics. 2016;10:1–11.
[39] Zhang P, Ren L, Zhang X, Shan Y, Wang Y, Ji Y, et al. Raman-activated cell sorting based on dielectrophoretic single-cell trap and release. Anal Chem. 2015;87:2282–9.
[40] McIlvenna D, Huang WE, Davison P, Glidle A, Cooper J, Yin HB. Continuous cell sorting in a flow based on single cell resonance Raman spectra. Lab Chip. 2016;16:1420–9.
[41] Cao C, Zhou D, Chen T, Streets AM, Huang YY. Label-free digital quantification of lipid droplets in single cells by stimulated raman microscopy on a microfluidic platform. Anal Chem. 2016;88:4931–9.
[42] Fan X, Zheng W, Singh DJ. Light scattering and surface plasmons on small spherical particles. Light: Sci Appl. 2014;3:e179.
[43] Redding B, Schwab MJ, Pan Y-L. Raman spectroscopy of optically trapped single biological micro-particles. Sensors (Basel, Switzerland). 2015;15:19021–46.
[44] Shoji T, Nohara R, Kitamura N, Tsuboi Y. A method for an approximate determination of a polymer-rich-domain concentration in phase-separated poly(N-isopropylacrylamide) aqueous solution by means of confocal Raman microspectroscopy combined with optical tweezers. Anal Chim Acta. 2015;854:118–21.
[45] Rkiouak L, Tang MJ, Camp JC, McGregor J, Watson IM, Cox RA, et al. Optical trapping and Raman spectroscopy of solid particles. PCCP. 2014;16:11426–34.
[46] Davidson N, Tong HJ, Kalberer M, Seville PC, Ward AD, Kuimova MK, et al. Measurement of the Raman spectra and hygroscopicity of four pharmaceutical aerosols as they travel from pressurised metered dose inhalers (pMDI) to a model lung. Int J Pharm. 2017;520:59–69.
[47] Lin M, Xu B, Yao H, Shen A, Hu J. An in vivo quantitative Raman-pH sensor of arterial blood based on laser trapping of erythrocytes. Analyst. 2016;141:3027–32.
[48] Lin D, Zheng Z, Wang Q, Huang H, Huang Z, Yu Y, et al. Label-free optical sensor based on red blood cells laser tweezers Raman spectroscopy analysis for ABO blood typing. Opt Express. 2016;24:24750–9.
[49] Ahlawat S, Chowdhury A, Uppal A, Kumar N, Gupta PK. Use of Raman optical tweezers for cell cycle analysis. Analyst. 2016;141:1339–46.
[50] Carney RP, Hazari S, Colquhoun M, Tran D, Hwang B, Mulligan MS, et al. Multispectral optical tweezers for biochemical fingerprinting of cd9-positive exosome subpopulations. Anal Chem. 2017;89:5357–63.

[51] Al Balushi AA, Kotnala A, Wheaton S, Gelfand RM, Rajashekara Y, Gordon R. Label-free free-solution nanoaperture optical tweezers for single molecule protein studies. Analyst. 2015;140:4760–78.
[52] Steven J, Ahmed AA, Reuven G. Raman spectroscopy of single nanoparticles in a double-nanohole optical tweezer system. J Opt. 2015;17:102001.
[53] Wu M-y, Ling D-x, Ling L, Li W, Li Y-q. Stable optical trapping and sensitive characterization of nanostructures using standing-wave Raman tweezers. Sci Rep. 2017;7:42930.

Oleg Ryabchykov, Shuxia Guo and Thomas Bocklitz

4 Analyzing Raman spectroscopic data

Abstract: This chapter is a short introduction into the data analysis pipeline, which is typically utilized to analyze Raman spectra. We empathized in the chapter that this data analysis pipeline must be tailored to the specific application of interest. Nevertheless, the tailored data analysis pipeline consists always of the same general procedures applied sequentially. The utilized procedures correct for artefacts, standardize the measured spectral data and translate the spectroscopic signals into higher level information. These computational procedures can be arranged into separate groups namely data pre-treatment, pre-processing and modeling. Thereby the pre-treatment aims to correct for non-sample-dependent artefacts, like cosmic spikes and contributions of the measurement device. The block of procedures, which needs to be applied next, is called pre-processing. This group consists of smoothing, baseline correction, normalization and dimension reduction. Thereafter, the analysis model is constructed and the performance of the models is evaluated. Every data analysis pipeline should be composed of procedures of these three groups and we describe every group in this chapter. After the description of data pre-treatment, pre-processing and modeling, we summarized trends in the analysis of Raman spectra namely model transfer approaches and data fusion. At the end of the chapter we tried to condense the whole chapter into guidelines for the analysis of Raman spectra.

Keywords: Raman spectroscopic data analysis, spectral preprocessing, spectral standardization, machine learning for spectral data, data analysis workflow

4.1 General analysis pipeline

Since the early 70s the potential of Raman spectroscopy for the characterization of biological samples like DNA, proteins and lipids was recognized. Nevertheless, it was until the 2000s that the technique could be utilized. One reason was the development of components needed for the application of Raman spectroscopy for biological samples (see chapter 2). The other reason is that powerful statistical and computational methods are needed in order to translate the Raman spectral signals

---

Oleg Ryabchykov and Shuxia Guo share the main authorship.

This article has previously been published in the journal *Physical Sciences Reviews*. Please cite as: Ryabchykov, O., Guo, S., Bocklitz, T. Analyzing Raman spectroscopic data *Physical Sciences Reviews* [Online] **2019**, 4. DOI: 10.1515/psr-2017-0043.

https://doi.org/10.1515/9783110515312-004

into meaningful bio-medical information. To do so powerful computers are needed, which can deal with large data sets. Additionally, tailored data analysis pipelines for the analysis of Raman spectra must be developed [1], which allow the application of Raman spectroscopy for real-world tasks. Possible applications include crystals and minerals (chapter 5), pharmacy (chapter 6), fine particles (chapter 7), medicine (chapter 8) archeological investigations (chapter 9) and forensics (chapter 10).

The data analysis pipeline must be tailored to the specific application of interest and is composed of computational procedures to correct for artefacts, standardize the measured spectral data and to translate the spectroscopic signals into higher level information. The procedures can be arranged in separate groups namely data pre-treatment, pre-processing and modeling. The pre-treatment aims to correct for non-sample-dependent artefacts, like spikes and contributions of the measurement device. This will be described in Section 4.2. The next block of procedures to be applied is called pre-processing and includes smoothing, baseline correction, normalization and dimension reduction. These methods are described in Section 4.3. Thereafter, the analysis model is constructed and evaluated, which will be described in Section 4.4. Every data analysis pipeline is composed of procedures of these three groups and the whole data pipeline to analyze Raman spectra is sketched in Figure 4.1. In Section 4.5 we will shortly describe new trends in the analysis of Raman spectra and in Section 4.6 a summary will be given, which focuses on dos and don'ts.

4.2 Data pretreatment

Like described earlier, the analysis of Raman spectra starts with a pre-treatment of the measured Raman spectra, which is necessary because the measured Raman spectra contain disturbing contributions and artefacts. These contributions corrupt the spectral information of the sample and prevent a reliable analysis. Thus, correction procedures need to be carried out. The most disturbing contributions within Raman spectra, which are not sample dependent, are cosmic spikes and contributions caused by experimental parameters and/or the measurement device itself. In order to deal with these contributions a spike correction, a wavenumber and intensity calibration are needed. These methods are described in the following subsections. The description starts with the spike correction, which should be carried out at the beginning of the data pipeline for Raman spectra. Thereafter a wavenumber calibration needs to be done and an optional intensity calibration might be performed.

4.2.1 Spike correction

In contrast to other corrupting effects in Raman spectroscopy, a presence of cosmic ray spikes in the spectroscopic data does not depend on the sample, laser, or

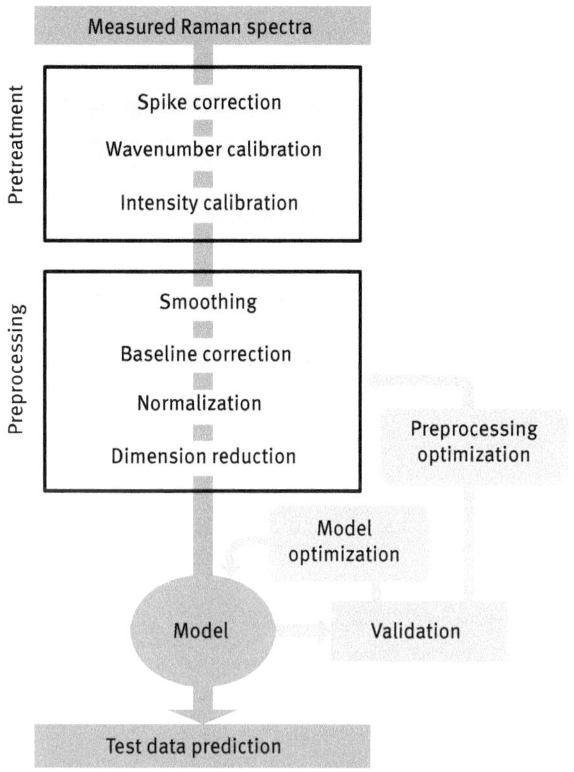

Figure 4.1: Data analysis pipeline for Raman spectra. A data pipeline used to analyze Raman spectroscopic data is shown. It is composed of pre-treatment, pre-processing and analysis procedures. The pretreatment steps remove corrupting effects which are not related to the sample and the preprocessing steps standardize the data by removing sample related contributions from the data. At the end of the pipeline, statistical models or machine learning approaches are constructed. These models are evaluated and there may be a parameter optimization based on the model outcome. All these steps aim a robust prediction of the constructed model.

spectrometer. Spikes appear at random positions in the data as values with large intensities (Figure 4.2). They occur when high-energetic cosmic particles hit the detector. In the CCD these particles generate electrons, which are read out along with the charges induced by the energy of Raman scattered photons. Therefore, cosmic particles introduce narrow features of high intensity, called spikes. Their positions are random and do not correspond to the wavenumbers directly. Typically, a spike appears within just one pixel or a few pixels.

The cosmic ray noise may affect the subsequent analysis, especially the normalization step. Therefore, many commercial devices perform multiple measurements in order to calculate an average spectrum. This approach applies when a small data set is acquired because it decreases all types of noise, including spikes. Unfortunately, it

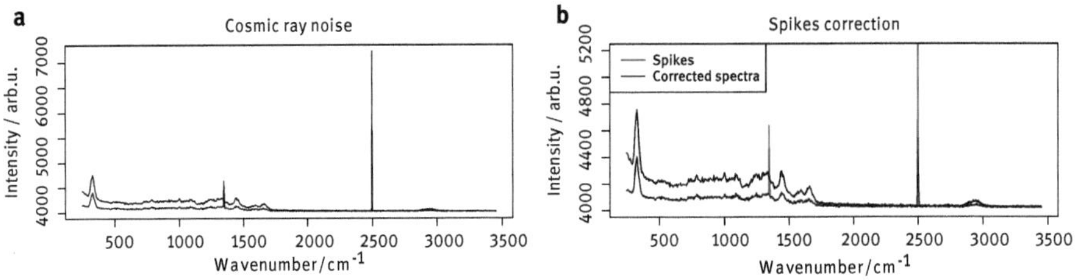

Figure 4.2: Cosmic ray noise. On the left side unprocessed Raman spectra containing spikes are shown. On the right side these spectra are shown in red and the corrected spectra are plotted over them in black. Therefore, only the spikes are visible in red color.

leads to the measurement time increase that is unsuitable for the acquisition of large spectral maps. Another method which works for small data sets is a manual removal of spikes found by visual inspection of the spectra. A drawback of manual spikes removal is that massive data sets cannot be processed by an operator within a reasonable time. To overcome this issue and automate the spike removal in massive data sets, specialized computational approaches were developed.

The simplest spike correction procedure is a median smoothing of the spectra. Unfortunately, besides spikes, this method filters and changes sharp spectral bands. Another option is a wavelet or Gabor transform with a suppressing of the coefficients corresponding to the spikes. Because spectral bands and spikes share the same frequencies, this approach can also corrupt the spectral information [2].

Besides being applied as filtering methods, wavelet or Gabor transform can be used to obtain a quantitative marker, related to the sharpness of features in the spectra. Then, spikes can be detected based on that marker and eliminated. Similarly, the marker can be obtained from nearest neighbors comparisons within the spectrum, or by applying a discrete Laplace operator [3]. If time series or scans are analyzed, the changes from spectrum to spectrum are typically small and the spikes detection methods can utilize the extra dimensions. Hence, 2-dimensional wavelets, comparison with the nearest neighbors within the spectral matrix or 2-dimensional Laplacian operator can be applied to enhance the reliability of the cosmic ray noise markers.

After extracting the quantitative marker, the spikes need to be located by setting a threshold. To estimate the threshold, the marker values can be compared to their standard deviation. For more robust comparison, the lowest and largest values may be excluded from the calculation of the standard deviation. Typically, a spike is considered to be present, if the response is higher than a preset threshold like three times the standard deviation. However, for some noise characteristics and sharpness of the Raman bands, this threshold may not be optimal. To optimize cosmic ray noise removal, the threshold should be selected depending on the distribution of the marker values [4].

4.2.2 Spectrometer calibration

The next step in the pre-treatment of Raman spectra is the spectrometer calibration. Ideally, the Raman spectroscopic signals should have high reproducibility and consistency. That means the measurement should be independent of the device and its instrumental configurations. In reality, however, a recorded Raman spectrum is affected by measurement conditions and does not solely reflect the sample. Variations in instruments, temperature, physical and chemical states of samples (e. g. solid or liquid) can lead to substantial spectral changes like wavenumber shifts and intensity variations [5, 6]. Furthermore, the instrumental response function changes over time. Raman spectra measured with a time delay can be different, even if they are measured on identical samples and under the same conditions. A calibration procedure is often required to reduce the introduced spectral changes, which includes wavenumber calibration and intensity calibration, as described in the following [5–7].

4.2.2.1 Wavenumber calibration

A CCD detector is typically used in a Raman spectrometer to collect Raman scattered photons at different frequencies (i. e. wavenumbers) with a different pixel (Figure 4.3a). Each spectrometer has a defined relation between wavenumber and pixel position [8]. However, the relation is sensitive to environmental and instrumental changes, like variations of temperature, the replacement of an instrumental component, or drifts of the instrument over-time. As a consequence, a CCD pixel can correspond to a different wavenumber. Hence the Raman spectrum is not recorded with the correct wavenumber axis, which manifests itself as wavenumber shifts compared to the theoretical values. This is shown by the two spectra in Figure 4.3a. The wavenumber shifts make it problematic to compare Raman spectra measured under different conditions or analyze them together.

A wavenumber calibration is conducted to find the correct wavenumber axis and thus remove the wavenumber shifts. To do so, a standard material with well-defined Raman bands is measured before measuring real samples. As shown in Figure 4.3b, the positions of these known Raman bands are located on the measured Raman spectrum. Thereafter the wavenumber differences between the observed and theoretical Raman bands are calculated. Based on these differences a parametric function is fitted and interpolated to all recorded wavenumber positions forming the wavenumber axis. Provided the Raman spectra of the standard material and the sample spectra share the same wavenumber axis, the wavenumber axis of the Raman spectra can be corrected by the obtained correction function, which removes the undesired wavenumber shifts.

The precision of wavenumber calibration is dependent on several factors [7]. Among those a careful selection of the standard material is highly important. The known Raman bands of the standard material need to be densely distributed and

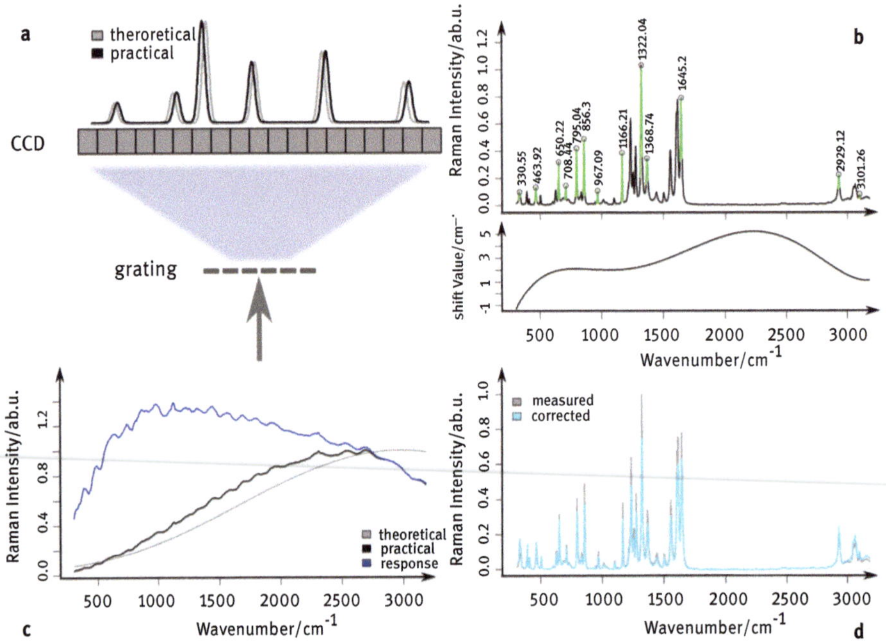

Figure 4.3: Workflow of wavenumber (a–b) and intensity calibration (c–d). (a) The relation between wavenumber and pixel positions can change, leading to wavenumber misalignment between measured and theoretical Raman spectra. (b) The wavenumber misalignment is corrected based on Raman spectra of a known standard material. (c) The intensity response function of the device is calculated as the ratio between measured and theoretical emission of a known standard material. (d) Intensity axis of measured Raman spectra is corrected by the calculated intensity response function.

cover the whole spectral range of interest. For biological applications the bands of the standard material should cover the fingerprint region and the CH-stretching region. Additionally, the number of Raman bands needs to be sufficient to stably interpolate the correction function to the whole spectral range. Materials that can be used for wavenumber calibration and their tabulated Raman bands are available in [9, 10]. One widely applied example in biological studies is 4-acetamedophenol (Figure 4.3b).

Other influential factors for a precise wavenumber calibration are the quality of the peak searching and the subsequent fitting of the correction function. For the first aspect, the wavenumber positions of tabled Raman bands are located by interrogating a neighborhood of a given Raman band. The result can be the peak point of the neighborhood or the peak point of a Gaussian or Lorenz curve fitted to this neighborhood. For the second aspect, the correction function is typically fitted as a polynomial with a degree of three to five. Polynomials with higher degrees are not recommended.

4.2.2.2 Intensity calibration

The intensity of a recorded Raman spectrum is the product of the true Raman scattered intensity (including baseline intensities) and the intensity response function of an instrument (Figure 4.3c) [7, 11]. In principle, quantitative and qualitative studies can be performed without considering this fact as long as the intensity response function remains unchanged over the measurements. However, the intensity response function does change with instrumental and environmental factors like excitation wavelength changes, detector replacement, changes of the sampling geometry, temperature, and so forth. The intensity response function of the same instrument can also change over time. Consequently, the intensities of recorded Raman spectra vary from instrument to instrument, condition to condition, and time to time. Such intensity variations can be ignored in most qualitative studies. For quantitative analyses and the comparison with a spectral library, the influence of the intensity variations becomes a critical issue and has to be corrected.

Therefore, intensity calibration is required in order to correct the recorded Raman spectrum with the intensity response function. Similar to wavenumber calibration, the intensity calibration also needs a standard material with known emission at different frequencies. The intensity response function is considered unchanged during the measurements of the standard material and actual samples, given they are measured under the same condition. As illustrated in Figure 4.3c and d, the procedure includes three steps. (1) The emission of a standard material is measured under the same condition as for the samples of interest. (2) The intensity response function is calculated as the ratio between the measured and theoretical emission of this standard material. (3) The Raman intensities of actual samples are corrected by dividing with the calculated intensity response function [5].

The efficiency of intensity calibration largely relies on the standard material. An ideal standard material should be homogeneous and give reproducible emission over broad wavenumber range. Existent standard materials (SRM) can be either a black-body radiator or a luminescence standard; both are available at NIST (National Institute of Standards & Technology, Gaithersburg, MD, USA) [12]. Black-body radiators are less often employed due to stability issues and difficulties to duplicate the sampling condition. The luminescence standards are more widely utilized in Raman spectroscopy. More details are beyond the scope of this chapter and interested readers are referred to [7, 11].

Above all, spectrometer calibration is proven to improve the results of the statistical analysis for datasets measured under different conditions, thanks to the improved spectral consistency over different measurements [13]. However, calibration cannot completely remove all undesired conditional relevant spectral variations [6, 14], which originates from multiple reasons including inaccessibility of a perfect standard material and unavoidable changes for measuring the standard material and the samples. The remaining spectral variations can still be disturbing for subsequent analysis and have to be handled. This leads to the topic of model transfer, which will be described in Section 4.5.1.

4.3 Data preprocessing

After the pre-treatment is carried out and artefacts of the measurement are corrected for, the pre-processing needs to be performed [15]. In this part sample related artifacts and sample dependent spectral contributions are corrected, leading to a standardization of the spectra. The most important correction procedure is a baseline correction, because the fluorescence background might be orders of magnitude stronger than the Raman signal. Before (or after) the baseline correction a smoothing can be carried out to correct noise contributions, but this is rather an optional step. Nevertheless, a few baseline correction procedures need a smoothed spectrum to construct a reliable baseline estimate. After these corrections are carried out, a normalization is performed to statistically standardize the spectra and a dimension reduction is done. The last step can also be done directly within the statistical model, e. g. some methods do this implicitly. Nevertheless, in principle, it is advisable to work with a lower dimensional representation of the spectra. In the following, we describe all parts of the pre-processing starting with smoothing procedures.

4.3.1 Smoothing

There are various types of noise that corrupt Raman spectra. They can be categorized into groups according to the source of the noise or according to the appearance in the data. However, as the cosmic ray noise standing out from the other types was already described in Section 4.2.1, we will not discuss it here.

In contrast to cosmic spike noise, the random noise in a Raman spectrum can be additive or intensity dependent. The additive noise has a Gaussian distribution and does not depend on the signal intensity. It corresponds mostly to the detector's dark current and readout noise. On the other side, the intensity dependent noise increases with increasing signal intensity and follows the Poisson distribution. To suppress this intensity dependent noise, it is important to plan the experiment in a way to keep the intensity of a fluorescence background low.

Although adjusted measurement conditions can minimize the noise, a completely noise-free spectrum cannot be measured. The noise in measured spectra can affect the baseline correction, normalization and detection of peak positions. To reduce the noise and make the interpretation of spectra easier, spectral smoothing can be applied.

Prominent smoothing procedures are Savitzky-Golay, mean, Gaussian, and median filtering. Each method has its own specifics. The Savitzky-Golay smoothing

[16], which is based on the least square fitting, is the most effective in preserving the peaks from corruption. On the other side, mean and Gaussian filters allow an efficient de-noising, and the median filter allows removing outliers from the spectrum. Although any filtering may remove parts of useful spectral information along with the noise, the corruption of the spectra can be avoided completely, if the size of the data set is large enough. In large data sets, a smooth appearance of spectra can be obtained by averaging over a large number of spectra. The influence of noise on the further analysis can also be reduced by a dimension reduction. To preserve the data from corruption, filtering should be avoided if there is no specific reason for filtering and large data sets are analyzed.

4.3.2 Background correction

There are two different types of baseline correction procedures in Raman spectroscopy: (1) subtracting the signal with the shutter closed from the spectrum and (2) subtracting the mathematically estimated baseline. For further discussion and differentiation between these both methods, the second method is referred to as a baseline correction. The baseline correction is of high importance for standardization of the spectral data when the samples feature a fluorescent background.

Estimating of the fluorescent background mathematically is based on the fact that fluorescence signal is broader than Raman spectral bands. Based on this property, a variety of algorithms for baseline correction were developed. Among the most typical ones (Figure 4.4a) we can highlight the modified polynomial fit [17], the asymmetric least squares baseline estimation [18], and the statistics-sensitive non-linear iterative peak-clipping (SNIP) algorithm [19]. The last one, in contrast to the others, does not lead to distortions of the baseline at the edges of the spectral

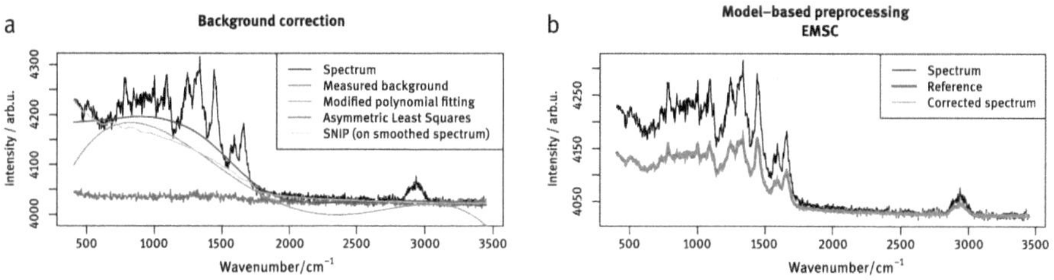

Figure 4.4: (a) Examples of background estimation. An example spectrum is shown in black color and the measured background signal is shown in red color. The other lines depict estimation of the fluorescence background by means of various baseline correction methods. (b) Model-based preprocessing. The spectrum after the data pretreatment is shown in green. For standardizing the spectra, an extended multiplicative scatter correction (EMSC) can be applied. The reference, which is typically an average spectrum over the data set, is depicted in red color, and the corrected spectrum is shown in green color.

interval. For a better performance of the SNIP baseline algorithm, the estimation should be carried out based on a smoothed spectrum. However, this estimated baseline can be subtracted from the non-smoothed spectrum, preserving the features that could be filtered out by smoothing.

If the variations of the Raman signals within the data set are expected to be small, a model-based preprocessing approach can be used. For example, an extended multiplicative scatter correction (EMSC) [20] is a powerful preprocessing tool that standardizes spectra according to chosen reference spectrum (Figure 4.4b). An additional advantage of this method is that further normalization is not required and the replicate variations within the data can be taken into account.

Both approaches, namely model-based preprocessing and baseline correction methods, should eliminate the background contribution of the spectra. Therefore, it is highly important to optimize their parameters based on the complexity of the background. Commonly the corrected spectra are investigated visually to estimate the goodness of the correction. A more robust approach is the introduction of a quantitative marker for the quality of baseline correction [21]. This marker should be based on expert knowledge about the spectroscopic data, such as regions where no background is expected and where the Raman spectroscopic signals should be located. If these regions influence the parameter differently, the correction approach, which features the extremum (maximum or minimum value) of the marker, would correspond to the optimal preprocessing.

4.3.3 Normalization

After the baseline correction, the Raman spectra become more standardized and in some cases can be analyzed directly. Unfortunately, the intensity variations of Raman spectra between investigated samples and even within spectral maps can be dramatic due to the changes in focusing and other experimental factors. An elimination of this effect is possible by applying a normalization step. Out of a huge range of normalization techniques, a few methods are commonly applied: vector normalization, normalization to the integrated spectral intensity, standard normal variate (SNV) and min-max scaling [22].

The vector normalization is performed by dividing the spectrum by the square root of the sum of the squared spectral intensities. It is conceptually similar to the root mean square normalization. This normalization can be also performed separately for different spectral regions, which could be needed in specific cases. Thus, in the analysis of biological samples, such as bacteria or fungi, better performance may be achieved by normalizing the fingerprint region and the CH-stretching region separately. As well as vector normalization, the normalization to the integrated spectral intensity, or area normalization, can be performed separately for different wavenumber regions. Besides that, the entire spectrum can be normalized to the intensity of a specific band, which is stable within the dataset. Furthermore, the l_1-normalization [23] is similar to the area normalization but operates with absolute values. So, in the case of l_1-norm, normalization to the integrated absolute spectral intensity is performed.

The next typical normalization approach is SNV scaling. It is performed by subtracting the mean intensity from the spectra and then dividing the result by the standard deviation of the spectrum. This method removes the constant background from the data. Thus, SNV is suitable to be applied without preliminary baseline correction in cases of a simple constant background.

Another scaling which eliminates the constant background is min-max normalization. It is performed by subtracting the minimum value of the spectrum and then dividing it by the maximum value of the resulting spectrum. This scaling approach is easy to use, but it is more sensitive to noise than other normalization methods.

Alternatively to normalization and scaling approaches, a model-based preprocessing, such as EMSC can be used, which was already mentioned in the Section 4.3.2 (Figure 4.4b). This approach does not require additional normalization because it combines both baseline correction and normalization.

4.3.4 Dimension reduction

Raman spectral datasets are mostly composed of a large number of variables, which poses challenges for a statistical analysis in terms of generalization performance as well as computational effort. A dimension reduction is needed to seek for a lower-dimensional representation of the original dataset without significantly losing key information [24]. The most straightforward way is to choose only the peak positions of Raman bands (probably together with their neighboring data points). This can be combined with a peak fitting procedure. However, this approach is not applicable if the Raman bands are unknown a priori, which is often the case. Hereby we briefly introduce more advanced approaches that have been widely applied.

Existent dimension reduction approaches can be categorized based on two properties. On the one hand, a lower-dimensional representation can be found with or without the presence of response variables such as group information or concentration of certain chemical components. The respective approaches are termed supervised and unsupervised dimension reduction approaches. On the other hand, the lower-dimensional representation can be based on the same variable space as the original data or based on a transformed space of the original data. Accordingly, the approaches are termed feature selection and feature extraction, respectively. A dimension reduction technique features both aspects, for instance, a supervised feature selection method.

Principal component analysis (PCA) is the most commonly applied unsupervised feature extraction method and the PCA model can be written as $\mathbf{X} = \mathbf{T}\mathbf{V}^T + \mathbf{e}$.

The original dataset $\mathbf{X} \in \mathbb{R}^{m,q}$ is mapped onto $r(r = \min(m, q))$ uncorrelated vectors V_j, namely the principal components (PC) or loadings. Each PC represents a different source of variances in **X**, with the largest variance in the first PC, the second largest in the second PC, and so forth. The calculation is achieved by a singular value decomposition on **X** or an Eigen value decomposition on the covariance matrix $(\mathbf{X}^T\mathbf{X})$. The loadings $V_j(n < j \leq r)$ are usually removed, because they mainly correspond to noise and are irrelevant to further analysis. In this way, the dataset **X** is represented by a lower dimensional score matrix $(\mathbf{T}^{m,n})$ and the corresponding error is denoted as **e**. Besides PCA, other unsupervised feature extraction approaches include independent component analysis and non-negative matrix factorization [25]. In particular, multivariate curve resolution alternating least squares (MCR-ALS) has shown its power in spectral analysis due to its capability of decomposing spectroscopic mixtures into multiple pure components and their concentrations. The concentration maxtrix can be used as scores for subsequent qualitative and quantitative analysis. By applying different constraints like non-negativity, unimodality and local rank, MCR-ALS can provide physically and chemically meaningful decomposition [26, 27]. Another commonly applied dimension reduction method is partial least squares (PLS) modeling. It is a supervised feature extraction method and bears some relation to PCA. Hereby the matrices of predictors $(\mathbf{X} \in \mathbb{R}^{m,q})$ and responses $(\mathbf{Y} \in \mathbb{R}^{m,p})$ are decomposed as $\mathbf{X} = \mathbf{T}\mathbf{P}^T + \mathbf{e}_1, \mathbf{Y} = \mathbf{U}\mathbf{Q}^T + \mathbf{e}_2$. The decomposition is performed so that the covariance between **T** and **U** is maximized. The dataset **X** is transferred into a lower-dimensional score matrix $\mathbf{T}^{m,n}(n < \min(m, q))$ by using the first n latent variables. All these described methods share the similarity that they decompose the observed dataset as a linear combination of N vectors and are called factor methods [28]. Other feature extraction methods like wavelet transform are also used in some investigations [29].

Unlike feature extraction, feature selection works by choosing variables that perform the best according to a predefined metric [24]. It has been well proven that a variable that is completely useless by itself can be significantly useful in combination with other variables. Hence a subset of variables is often selected simultaneously in practice. Approaches of subset selection include three categories: wrapper, filtering, and embedded methods. With wrapper methods, the optimal feature subset is selected to obtain the best prediction on data independent of the training data. Wrapper methods are computationally expensive due to the requirement of model training. Filter methods select a feature subset according to a certain metric that is independent of subsequently applied statistical models, for example, mutual information, Pearson's correlation coefficient, Fisher's discriminant ratio, or results from statistical tests. Filter methods are advantageous in terms of computational cost but they are less powerful to build a good predictive model. Nonetheless, filter methods can be used as a pre-selection prior wrapper methods. Embedded methods conduct feature selection as a part of model construction. This can be achieved by enforcing most coefficients of the model to be zero, like in the cases of LASSO [30] and sparse PLS [31]. In addition, feature selection can also be performed based on the variable weights/importance given by statistical models like support vector machine and random forest (RF). In all cases of feature selection approaches, a search procedure for feature subsets has to be utilized, be it genetic algorithms, simulated annealing or greedy search (i. e. forward/backward feature selection). For wrapper and embedded approaches, a way of assessing the prediction performance has to be known as well, which overlaps with the issue of model selection and is outlined in the following section.

Besides typical factor methods and feature selection, it is worth to note that nonlinear dimension reduction approaches have also been reported, for example, Isomap [32], locally linear embedding (LLE) [33, 34], feature learning with auto-encoder and other neural network framework [35, 36].

A common question in terms of dimension reduction is to find the best dimension, which typically refers to the optimal number of components in feature extraction or the best feature subset in feature selection. This task leads to the topic of model selection/optimization. In practice, it is achieved by searching for a trade-off between error and variance [37]. That is to say, to find a compromise between the error on the training data as well as a good generalization performance on test data. A routine procedure is to split the available dataset into training and validation data and optimize the model by minimizing the prediction error of the validation data. Figure 4.5 illustrates an example of dimension reduction conducted with a factor method, where the original dataset ($\mathbb{R}^{k\times m}$) is reduced to $\mathbb{R}^{k\times d}$. The training and validation data is represented by blue and green blocks, respectively. Dimension reduction and statistical modeling are conducted with the training data only. The validation data is predicted afterward. The performance of the prediction is benchmarked by a pre-determined metric. To get a more stable optimization, a cross-validation is often applied [38], where the dataset is re-split into training and validation datasets for several times. The metric of the prediction is averaged over the splits to determine the optimal dimension (d).

A crucial issue to find the best dimension is that the optimal result of dimension reduction is dependent on the model applied subsequently. Therefore, the optimization has to be conducted for dimension reduction and the subsequent model altogether, like in Figure 4.5. Another critical issue is that after the optimization, an additional validation is necessary to evaluate the performance of the optimized model, namely, the external validation. The normal way is to predict data that is not used as training or validation data (red block in Figure 4.5). In this case, the unknown data must be excluded during the model construction and optimization, especially if supervised dimension reduction methods are employed [39]. More details on this topic can be found in Section 4.4.3 of this chapter.

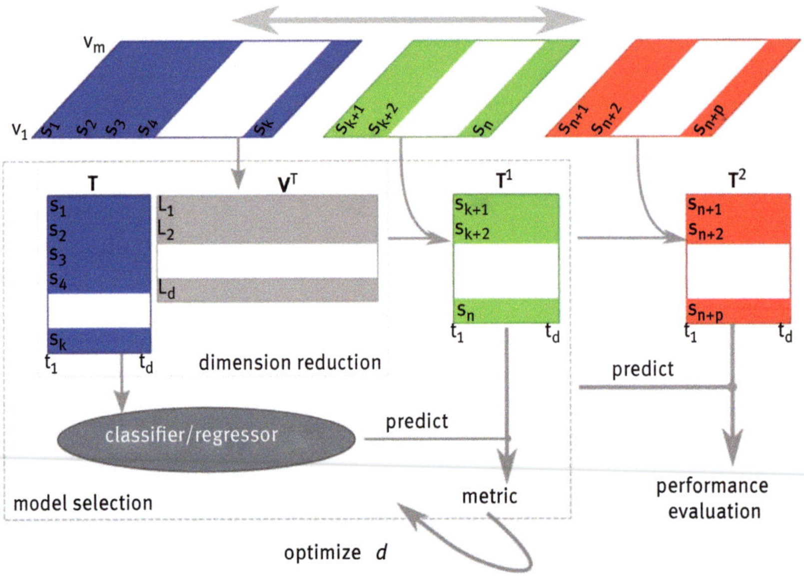

Figure 4.5: Workflow of dimension reduction and statistical modeling. The dimension reduction can be done with factor methods like PCA. The three data blocks shown in blue, green, and red represent training, validation, and testing dataset, respectively. The training dataset is used to build the model. Parameters of the model (such as the number of principal components of the PCA, d) are optimized based on the prediction on the validation dataset (green). Afterwards, the model is evaluated according to the prediction on the testing dataset (red).

4.4 Models

After the Raman spectra are pretreated and preprocessed, statistical models are applied in order to extract high-level information, such as concentrations of substances, distribution of substances, disease markers or sample types. The so-called statistical methods aim to translate the standardized Raman spectra into high-level information of interest, which can be further used by chemists, biologists and physicians. As most of these methods have a statistical origin we call them statistical methods even though a few are developed within the framework of machine learning [40].

The statistical methods applied for the analysis of Raman spectra are standard techniques and we group them according to their application scenario. We will first introduce clustering algorithms and unmixing procedures utilized for image generation. It should be noted here that the methods for dimension reduction can be utilized for the image generation as well. Thereafter we describe supervised methods including classification models for diagnostics and regression procedures for analytics [1].

4.4.1 Clustering and unmixing for imaging

There are two major clustering algorithm types: hard clustering and fuzzy clustering. In the former a spectrum belongs to one certain cluster and no any other clusters. The latter methods are related to unmixing. A spectrum belongs to multiple clusters to a certain extent, which is called cluster membership. Both types of clustering methods are widely used in Raman spectroscopy, especially for imaging purposes to produce an overview. The most often applied clustering algorithms are *k*-means clustering and hierarchical clustering [41]. The *k*-means clustering starts with a random cluster distribution of *k* clusters. Then the distance of all spectra to the cluster mean spectra are evaluated and the spectra are resorted corresponding to the minimal distance. The procedure should be performed for multiple times because the algorithm is greedy. The most common version of hierarchical clustering is the agglomerative clustering, where Raman spectra are merged to clusters until only one cluster exists [40]. As described above there are also fuzzy clustering versions like fuzzy c-means clustering [42], which is the extension of the hard *k*-means clustering. If the task is to determine mixture compositions, unmixing methods are the ideal tools. To extract pure components from the data without a training dataset the so-called end-member extraction methods were developed. They extract the most "extreme" spectra in a specific sense from the data. Methods that are commonly applied for end-member extraction are N-FINDR [41] and Vertex Component Analysis (VCA) [41]. Besides these techniques the multivariate curve resolution-alternating least squares (MCR-ALS) method gains more and more attention due to the incorporation of different constraints and additional knowledge about the data [27].

4.4.2 Classification for diagnostics and regression models for analytics

If no training data with reference values are available, clustering or unmixing are the only techniques which can be applied. If training data with reference values are available, supervised machine learning algorithms, like regression or classification models, can be applied to extract high-level information [43]. Linear classification and regression methods are frequently applied due to their simplicity and robustness. Even though linear models often perform well, in many cases more powerful techniques are needed. Among those the most often utilized ones are kernel support vector machines (SVMs) and artificial neural networks [43]. Especially deep artificial neural network are powerful emerging techniques for classification and regression [44]. Another powerful classification and regression algorithm is the random forest (RF) model, which is an ensemble based method [45]. RF constructs a pre-defined number of random decision trees and every tree is able to predict. The output of the whole RF is generated by a voting procedure at the end. While certain models like SVMs or ANNs can be used intrinsically as classification and regression model, pure regression models can be converted to classification models using pseudo-concentrations. Among the most often applied regression models are principal component regression and PLS regression [46].

4.4.3 Evaluation procedures

A statistical model with perfect predictive performance is only possible if the training data is a complete representation of the population under investigation. However, this is usually not the case in real-world tasks and the data at hand is always a limited sampling of the population. The property of the population has to be estimated from the (limited) sampled data. In terms of chemometrics, the estimation refers to constructing a statistical model for a given classification or regression task. Due to the incomplete sampling, errors almost always occur when predicting new data with a trained model. In extreme cases, the model may fit the training data perfectly but cannot be generalized to unknown data. This is termed overfitting. To avoid overfitting, a procedure is required to evaluate the generalization performance of the model before it can be used in practice. To do so, the model is used to predict data that has not been used during model construction [37, 47]. The performance of the prediction can be benchmarked by different metrics, as is described in the following.

For the evaluation of regression models the prediction performance is measured as differences between the true and predicted values, for instance, the root-mean-squared-error of prediction (RMSEP; $\mathrm{RMSEP} = \sqrt{1/N \sum_{i=1}^{N} (\hat{y}_i - y_i)^2)}$ or the mean-squared-error of prediction (MSEP; $\mathrm{MSEP} = \mathrm{RMSEP}^2$). To quantify the performance of classification/clustering models, the metrics can be used are accuracy, sensitivity, specificity and Cohen's kappa. These values are derived from a confusion matrix, which is the combination of the predicted and true group assignments (see Table 4.1). A model can be evaluated with a combination of multiple metrics according to certain fusion regimes like the sum of ranking difference (SRD) [48]. This combination can provide more stable and reliable model evaluations than using a single metric alone.

As shown in Figure 4.5, the model evaluation requires a prediction of new data that is independent of the training data. This requires additional samples to be measured, which can be expensive, time-consuming, or even impossible. A more practical solution is resampling, where the datasets for model training and prediction are independently sampled from the accessible data. The most widely applied resampling regime is cross-validation [49]. Thereby the accessible dataset is split into training and testing data, used for model training and evaluation, respectively. The splitting procedure is repeated for several times to get a stable validation. The

Table 4.1: Confusion table. The table compares the prediction against the correct group assignment. From this table a number of classification characteristics, like accuracy, sensitivity and specificity, can be calculated.

		Predicted	
		P	¯P
True	P	**a**	b
	¯P	c	**d**

Accuracy $= 100\% \times \frac{a+d}{a+b+c+d}$, percentage of correctly classified samples.
Sensitivity $= 100\% \times \frac{a}{a+b}$, percentage of true positive.
Specificity $= 100\% \times \frac{d}{c+d}$, percentage of true negative.
$\kappa = \frac{\text{Accuracy} - p_e}{1 - p_e}$, agreement between truth and prediction.

results of the prediction over all splits are averaged to benchmark the generalization performance of the statistical model. According to the data splitting scheme, cross-validation can be conducted in different ways, including leave-p-out cross-validation, k-fold cross-validation, and Monte Carlo cross-validation [50]. A special case of cross-validation is holdout validation, where the data split is done only once without repetition. Another important resampling method is bootstrapping, which is a resampling procedure with replacement. No matter how data split is performed, the proportional composition of classes (or concentrations) in every split should be consistent to the composition of the population, because the constructed model can be influenced by the relative compositions of different classes in the training data. One way to achieve this is the Latin-partition method [51]. Specifically, bootstrapping with Latin partition was reported, which constructs multiple Latin partitions with a bootstrap. This allows getting the relation between the prediction and the composition of the training data as well as the optimization of the statistical model [52, 53].

There are two issues extremely crucial to model evaluation [38, 39, 54]. The first issue is the independence requirement between training and testing data. In practice, this is ensured by resampling the data on the highest level of sampling hierarchy, which might be the biological replicate, cultivation, or patient. With k-fold cross-validation, for example, the folds should be arranged according to the highest level of sampling hierarchy. Otherwise, the information of testing data is implicitly used during model construction and the prediction on the testing data does not reveal the true generalization performance. As a result, the statistical model is overestimated. In addition, the evaluation should be conducted involving both the dimension reduction and the statistical model. This is especially important if a supervised dimension reduction is applied. In special cases, if parameters of a pre-processing procedure are optimized according to the output of the statistical models, the evaluation loop should include the pre-processing steps as well.

Noteworthy, it requires special attention when both the model selection and model evaluation are conducted with cross-validation. In this case, a two-layer cross-validation is needed [39]. As shown in Figure 4.5, the dataset is split into training (blue and green blocks) and testing data (red block) within the outer-layer cross-validation. The testing data is taken aside and the training data is fed into the inner-layer cross-validation for model optimization as described in Section 4.3.4. Thereafter the testing data is predicted with the optimized model, of which the results are used to evaluate the performance of the model. The inner-layer and outer-layer cross-validation are termed internal and external validation, respectively.

4.5 New trends

As almost all applications of Raman spectroscopy are only possible, if an adequate data analysis pipeline is utilized, the research area developing new analysis methods and tools is active. A full summary of trends is beyond the scope of this chapter, but two topics, which emerged recently, should be summarized here. These both topics are model transfer and data fusion. Model transfer is dealing with the use of models, which were constructed based on training data different from the test data. Such a model transfer issue arises, if a chemometric model should be utilized for diagnostic purposes and the device on which the test data are measured in clinics is different compared to the device utilized for measuring the training data. This issue is important because it is linked to real-world applications of Raman spectroscopy. The next active research area is related to data fusion, where Raman spectra are computationally combined with other data types in order to extract more information as it would be possible from the Raman spectra alone. In that manner, complementary information to the information extracted from Raman spectra can be analyzed together with the Raman based information.

4.5.1 Model transfer

A well-known challenge in chemometrics is the substantially inferior prediction quality of a pre-trained chemometric model if it is applied to newly measured data [55]. This issue gets more important if the new data is significantly different to training data. In Raman spectroscopy, such differences manifest itself as wavenumber shifts and intensity variations (Figure 4.6a). One of the major reasons for such spectral deviations is the instrumental change over-time or after replacement of a component. The wavenumber/intensity calibration (see Section 4.2) helps to reduce such instrument induced spectral variations but cannot completely remove them. The remaining spectral variations can still mask the spectral differences of interest and thus corrupt the prediction, which is very common in biological

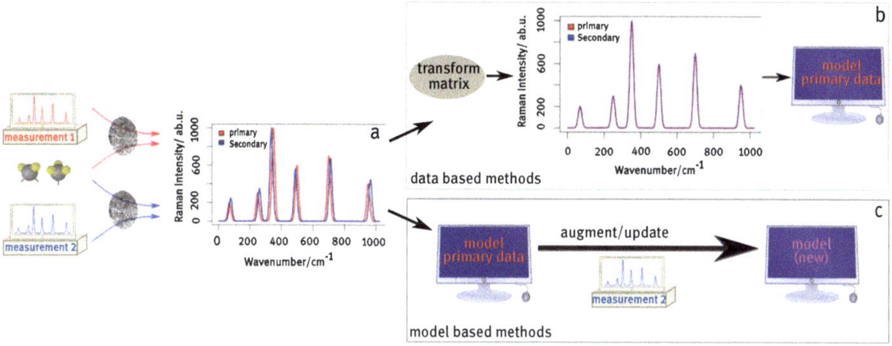

Figure 4.6: Overview of model transfer. (a) Training (primary) and testing (secondary) datasets can be significantly different if they are measured from different replicates or on different devices. Hence the chemometric model constructed with the primary dataset can fail to predict the secondary dataset. This can be tackled with model transfer approaches according to two mechanisms: data based methods (b) and model based methods (c).

studies. Besides the instrument variations, other experimental changes can also disturb the reproducibility, for example, cultivation conditions cannot be exactly identical for all replicates and differences over measurement of different replicates are resulting. These spectral variations cannot be tackled with calibration at all. That is why an existent model cannot successfully predict the newly measured data. A simple but labor-extensive solution is to train another model for this new data. However, this is not possible if new training samples are inaccessible, which might be the case in disease diagnosis. Therefore, a method is needed to enable the prediction of the new data based on the existent model. This is achieved with model transfer approaches, as described in the following [14, 56, 57].

In the terminology of model transfer, the (old) training and the new data are termed primary and secondary data, respectively. There are two types of model transfer approaches: data based (Figure 4.6b) and model based methods (Figure 4.6c) [58]. In the former case, the primary and secondary data are transformed to make them more similar. In the latter case, an existent model is updated to improve the prediction on the secondary data. Model transfer can be applied in a supervised or an unsupervised manner. Unsupervised model transfer is conducted without the knowledge of response variables (class information or concentration) of the secondary data. For supervised model transfer, a few secondary samples with known responses are needed; but the required sample size is much smaller compared to the construction of a new model. Unsupervised methods do not need the response information of the secondary data, making them superior to supervised model transfer in the cases where the response information is not accessible. A typical example of this case is bio-medical diagnostics, where the disease level of a new patient should be predicted and is unknown.

Data based model transfer aims to remove the spectral variations between secondary and primary data. Typically applied methods are Procrustes analysis and piecewise direct standardization (PDS) [59], where a transformation matrix is calculated to map the secondary to the primary data. Other approaches include warping methods, which adjust the peak misalignment between primary and secondary data [60]. These methods are typically conducted based on spectra of standard samples measured under primary and secondary conditions. Then the resulting correction function is applied to the spectra of the secondary samples. The bottleneck of these methods is that the standard have to be measured under both conditions. In addition, the correction function only corrects the instrumental deviations and cannot tackle disturbing effects like the variation of cultivation conditions. Hence the secondary and primary data of real samples cannot be perfectly matched. Another option is to calculate the transformation matrix based on the primary and secondary data of real samples itself. However, this works only if the samples for primary and secondary measurements share the same chemical components. In classification and regression tasks, the samples belong to multiple classes or feature different concentrations of their components. It is almost impossible to ensure that both the primary and the secondary samples feature the same classes or identical concentrations. In this case, the transformation matrix does not only model the undesired changes but also the spectral differences of interest. It is advisable to calculate transformation matrix separately for each class or concentration, i. e. in a supervised manner.

Model based model transfer seeks for a compromise between the primary and secondary data. The first scheme is to build a global statistical model involving experimental variables responsible for undesired spectral changes [58]. This procedure requires to know and to include all potential influential experimental variables, which makes it less feasible for model transfer. An alternative regime is to build a model on the primary data with features robust to the experimental changes. In ref [61] the authors proposed a sample-wise spectral multivariate calibration approach by penalizing and desensitizing features that strongly differ between the primary and secondary data. This method is less powerful if the secondary conditions differ strongly from the primary conditions. The third scheme is local modeling, where the model is built only with the primary samples that are the nearest neighbors to the secondary data [62]. It is crucial in local modeling to determine the number of nearest neighbors and the similarity metric to select the nearest neighbors. Another model transfer approach is model augmentation. Thereby, the training dataset is enlarged with several additional secondary samples, the so-called transfer samples [63].

So far model transfer has been mostly investigated for near-infrared spectroscopy and regression problems. Model transfer of Raman spectroscopy and classification tasks is rather new and only a few studies exist. Recently, a model transfer approach was developed for Raman spectroscopy using Tikhonov regularization based partial least squares regression (TR-PLSR) [14]. However, the method does not work if the response variables of the secondary data are unavailable. To deal with this issue, unsupervised model transfer approaches for Raman spectroscopy were also developed recently [57].

4.5.2 Data fusion

In cases when the Raman data does not yield sufficient information, it can be complemented by additional measured data. For example, if correlated imaging is performed, several types of spectroscopic or spectrometric data are measured and can be combined. Another example is that beside the Raman spectra of the sample other additional information can be used, like the patient's laboratory values, gender, age, physical parameters, and known medical conditions. These values can be used along with Raman spectra to improve the performance of a chemometric model. Often the different data types feature different dynamic range, dimensionality, and the number of observations per sample. Therefore, the question arises how different data types can be combined [64]. This process of combining different types of data is called data fusion.

The data fusion can be performed on different levels [65] of the analysis pipeline (Figure 4.7). The combination can be done on a low-level directly after the preprocessing, possibly even before the dimension reduction. This data fusion scheme is called centralized data fusion. It is performed by merging the data from different sources into a single data matrix with subsequent simultaneous analysis. However, dealing with different data types within a low-level data fusion approach requires accounting for different scaling, dynamic range and dimensionality of the data in order to balance the contributions of different data types against each other. To account for these differences, it may be important to rescale the data before combining them [66].

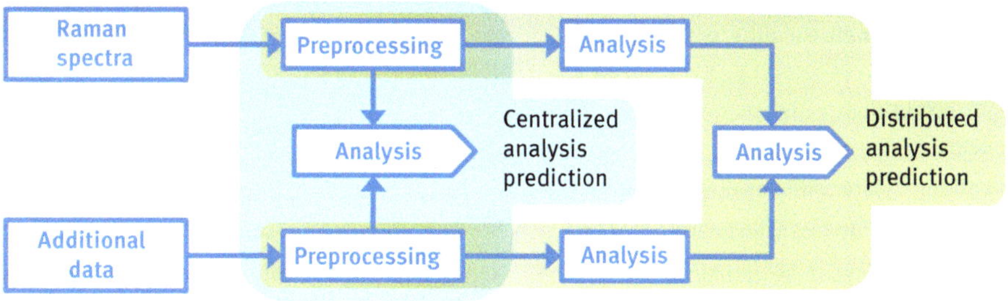

Figure 4.7: Schematic representation of data combination with a fusion center at different levels of the data analysis pipeline. The low-level (centralized) data fusion is highlighted in blue color, and the high-level (distributed) data fusion is highlighted in green color.

Another possible data fusion scheme is a high-level data fusion, also called a distributed data fusion. In this scheme, each data type is analyzed separately and the scores are combined at the final step of the analysis [67]. The advantage of the high-level data fusion is that it is computationally less costly and allows dealing with the different data types in an easier manner.

Besides the low-level and high-level data fusion, a decentralized data fusion approach [68] or hierarchical data fusion can be used. For example, in correlated imaging, the hierarchical data fusion allows using one imaging technique for finding areas of the interest. The other imaging technique is then utilized to study these regions [69, 70]. In that manner, deeper insights into the investigated areas can be gathered.

4.6 Summary: dos and don'ts in analyzing Raman spectra

In this section we would like to summarize the sections above. We would like to give the summary in terms of a Dos and Don'ts list. In that manner, we tried to condense the content of the whole chapter into guidelines and rules. In order to allow a further reading, the recommended practices discussed in the sections about data pretreatment

Do's	Dont's
Avoid fluorescence if possible (3) Check instrumental drift every day before measurement (2)	
Measure standard material for calibration every time before measuring real samples (2)	Measure a spectrum of standard material for calibration at different conditions (days) compared to real samples (2)
Wavenumber calibration and/or intensity calibration (2)	
	Direct application of a model to data of another device (2)
Model transfer between different devices/replicates (4)	Model transfer between datasets measured from different classes (4)
Apply baseline correction methods prior modeling (3)	
Smoothing before SNIP baseline correction (3)	
Normalization (after baseline correction) (3)	Normalization before baseline correction (3)
Involve procedures to be optimized like dimension reduction (pre-processing, if necessary) inside the CV loop (4)	Perform dimension reduction outside of CV loop, especially for supervised dimension reduction approaches (4)
Evaluate the model with independent dataset (e. g. external CV) (4)	Evaluate the model with data already used during model construction/optimization (4)
Use data from the same sample (replicate) exclusively as training or testing data (4)	Split the data into training and testing data regardless of the replicate information (4)
Leave-one-sample (replicate)-out CV (4)	Leave-one-spectrum-out CV (4)

(Section 4.2), data preprocessing (Section 4.3) and models (Section 4.4) are marked in the table with a respective chapter number in the parentheses.

References

[1] Bocklitz TW, Guo S, Ryabchykov O, Vogler N, Popp J. Raman based molecular imaging and analytics: a magic bullet for biomedical applications!? Anal Chem. 2016;88:133–51.

[2] Ehrentreich F, Sümmchen L. Spike removal and denoising of Raman spectra by wavelet transform methods. Anal Chem. 2001;73:4364–73.

[3] Schulze HG, Turner RF. A two-dimensionally coincident second difference cosmic ray spike removal method for the fully automated processing of Raman spectra. Appl Spectrosc. 2014;68:185–91.

[4] Ryabchykov O, Bocklitz T, Ramoji A, Neugebauer U, Foerster M, Kroegel C, et al. Automatization of spike correction in Raman spectra of biological samples. Chemometrics Intell Lab Syst. 2016;155:1–6.

[5] Dörfer T, Bocklitz T, Tarcea N, Schmitt M, Popp J. Checking and improving calibration of Raman spectra using chemometric approaches. Z Phys Chem Int J Res Phy Chem Chem Phy. 2011;225:753–64.

[6] Bocklitz T, Dörfer T, Heinke R, Schmitt M, Popp J. Spectrometer calibration protocol for Raman spectra recorded with different excitation wavelengths. Spectrochimica Acta A: Mol Biomol Spectrosc. 2015;149:544–9.

[7] McCreery RL. Raman spectroscopy for chemical analysis Vol. 157. New York: John Wiley & Sons, 2000

[8] Berg RW, Nørbygaard T. Wavenumber calibration of CCD detector Raman spectrometers controlled by a sinus arm drive. Appl Spectrosc Rev. 2006;41:165–83.

[9] Carrabba MM. Wavenumber standards for Raman spectrometry. In: Griffiths P, Chalmers JM, editor(s). Handbook of vibrational spectroscopy. Chichester: Wiley online library, 2006

[10] E1840-96, A., Standard guide for Raman shift standards for spectrometer calibration. ASTM International, West Conshohocken, PA, 2014. 03.06.

[11] Fryling M, Frank CJ, McCreery RL. Intensity calibration and sensitivity comparisons for CCD/Raman spectrometers. Appl Spectrosc. 1993;47:1965–74.

[12] Davis W, Forney G, Bukowski R. National institute of standards and technology, Gaithersburg MD, USA.

[13] Rodriguez JD, Westenberger BJ, Buhse LF, Kauffman JF. Standardization of Raman spectra for transfer of spectral libraries across different instruments. Analyst. 2011;136:4232–40.

[14] Guo S, Heinke R, Stöckel S, Rösch P, Bocklitz T, Popp J. Towards an improvement of model transferability for Raman spectroscopy in biological applications. Vib Spectrosc. 2017;91:111–8.

[15] Bocklitz T, Walter A, Hartmann K, Rosch P, Popp J. How to pre-process Raman spectra for reliable and stable models? Anal Chim Acta. 2011;704:47–56.

[16] Savitzky A, Golay MJE. Smoothing and differentiation of data by simplified least squares procedures. Anal Chem. 1964;36:1627–39.

[17] Lieber CA, Mahadevan-Jansen A. Automated method for subtraction of fluorescence from biological Raman spectra. Appl Spectrosc. 2003;57:1363–7.

[18] Eilers PH, Boelens HF. Baseline correction with asymmetric least squares smoothing. Leiden Univ Med Centre Rep. 2005;1:1.

[19] Ryan CG, Clayton E, Griffin WL, Sie SH, Cousens DR. SNIP, a statistics-sensitive background treatment for the quantitative analysis of PIXE spectra in geoscience applications. Nucl Instrum Methods Phys Res B: Beam Interact Mater Atoms. 1988;34:396–402.
[20] Martens H, Stark E. Extended multiplicative signal correction and spectral interference subtraction: new preprocessing methods for near infrared spectroscopy. J Pharm Biomed Anal. 1991;9:625–35.
[21] Guo S, Bocklitz T, Popp J. Optimization of Raman-spectrum baseline correction in biological application. Analyst. 2016;141:2396–404.
[22] Gautam R, Vanga S, Ariese F, Umapathy S. Review of multidimensional data processing approaches for Raman and infrared spectroscopy. EPJ Tech Instrum. 2015;2:8.
[23] Black MJ, Anandan P. The robust estimation of multiple motions: parametric and piecewise-smooth flow fields. Comput Vis Image Understand. 1996;63:75–104.
[24] Guyon I, Elisseeff A. An introduction to variable and feature selection. J Mach Learn Res. 2003;3:1157–82.
[25] Malinowski ER. Factor analysis in chemistry, 2 ed. New York: Wiley, 1991
[26] Zhang X, Tauler R. Application of multivariate curve resolution alternating least squares (MCR-ALS) to remote sensing hyperspectral imaging. Anal Chim Acta. 2013;762:25–38.
[27] Piqueras S, Krafft C, Beleites C, Egodage K, Von Eggeling F, Guntinas-Lichius O, et al. Combining multiset resolution and segmentation for hyperspectral image analysis of biological tissues. Anal Chim Acta. 2015;881:24–36.
[28] Brereton RG, Jansen J, Lopes J, Marini F, Pomerantsev A, Rodionova O, et al. Chemometrics in analytical chemistry – Part I: history, experimental design and data analysis tools. Anal Bioanal Chem. 2017;409:5891–9.
[29] Bruce LM, Koger CH, Li J. Dimensionality reduction of hyperspectral data using discrete wavelet transform feature extraction. IEEE Trans Geosci Remote Sens. 2002;40:2331–8.
[30] Tibshirani R. Regression shrinkage and selection via the Lasso. J R Stat Soc Series B Methodol. 1996;58:267–88.
[31] Chun H, Keleş S. Sparse partial least squares regression for simultaneous dimension reduction and variable selection. J R Stat Soc Series B Stat Methodol. 2010;72:3–25.
[32] Zhang Z, Chow TW, Zhao M. M-Isomap: orthogonal constrained marginal isomap for nonlinear dimensionality reduction. IEEE Trans Syst Man Cybern. 2013;43:180–91.
[33] Silva VD, Tenenbaum JB. Global versus local methods in nonlinear dimensionality reduction. In: Advances in neural information processing systems. Cambridge, MA, USA: MIT Press, 2003.
[34] Shan R, Cai W, Shao X. Variable selection based on locally linear embedding mapping for near-infrared spectral analysis. Chemometrics Intell Lab Syst. 2014;131:31–6.
[35] Hinton GE, Salakhutdinov RR. Reducing the dimensionality of data with neural networks. Science. 2006;313:504–7.
[36] Wang W, Huang Y, Wang Y, Wang L. Generalized autoencoder: a neural network framework for dimensionality reduction. In: Proceedings of the IEEE conference on computer vision and pattern recognition workshops. 2014.
[37] Kalivas JH, Palmer J. Characterizing multivariate calibration tradeoffs (bias, variance, selectivity, and sensitivity) to select model tuning parameters. J Chemom. 2014;28:347–57.
[38] Arlot S, Celisse A. A survey of cross-validation procedures for model selection. Stat Surv. 2010;4:40–79.
[39] Guo S, Bocklitz T, Neugebauer U, Popp J. Common mistakes in cross-validating classification models. Anal Methods. 2017;9:4410–7.
[40] Hastie T, Tibshirani R, Friedman J. The elements of statistical learning; data mining, inference and prediction. New York: Springer, 2008.

[41] Hedegaard M, Matthäus C, Hassing S, Krafft C, Diem M, Popp J. Spectral unmixing and clustering algorithms for assessment of single cells by Raman microscopic imaging. Theor Chem Acc. 2011;130:1249–60.
[42] Bezdek JC, Ehrlich R, Full W. FCM: the fuzzy c-means clustering algorithm. Comput Geosci. 1984;10:191–203.
[43] Bocklitz T, Putsche M, Stüber C, Käs J, Niendorf A, Rösch P, et al. A comprehensive study of classification methods for medical diagnosis. J Raman Spectrosc. 2009;40:1759–65.
[44] Acquarelli J, Van Laarhoven T, Gerretzen J, Tran TN, Buydens LM, Marchiori E. Convolutional neural networks for vibrational spectroscopic data analysis. Anal Chim Acta. 2017;954:22–31.
[45] Breiman L. Random forests. Mach Learn. 2001;45:5–32.
[46] Mevik B-H, Wehrens R, Liland KH. pls: partial least squares and principal component regression. R Package Version. 2011;2(3).
[47] Vehtari A, Gelman A, Gabry J. Practical Bayesian model evaluation using leave-one-out cross-validation and WAIC. Stat Comput. 2017;27:1413–32.
[48] Héberger K. Sum of ranking differences compares methods or models fairly. TrAC Trends Anal Chem. 2010;29:101–9.
[49] Refaeilzadeh P, Tang L, Liu H. Cross-validation, in Encyclopedia of database systems. New York: Springer, 2009:532–8.
[50] Xu QS, Liang YZ, Du YP. Monte Carlo cross-validation for selecting a model and estimating the prediction error in multivariate calibration. J Chemom. 2004;18:112–20.
[51] Wan C, Harrington PDB. Screening GC-MS data for carbamate pesticides with temperature-constrained-cascade correlation neural networks. Anal Chim Acta. 2000;408:1–12.
[52] De Boves Harrington P. Statistical validation of classification and calibration models using bootstrapped Latin partitions. TrAC Trends Anal Chem. 2006;25:1112–24.
[53] Qi N, Zhang Z, Xiang Y, Yang Y, Liang X, Harrington PD. Terahertz time-domain spectroscopy combined with support vector machines and partial least squares-discriminant analysis applied for the diagnosis of cervical carcinoma. Anal Methods. 2015;7:2333–8.
[54] Krstajic D, Buturovic LJ, Leahy DE, Thomas S. Cross-validation pitfalls when selecting and assessing regression and classification models. J Cheminform. 2014;6:10.
[55] Copas JB. Regression, prediction and shrinkage. J R Stat Soc Series B Methodol. 1983;45:311–54.
[56] Shahbazikhah P, Kalivas JH. A consensus modeling approach to update a spectroscopic calibration. Chemometrics Intell Lab Syst. 2013;120:142–53.
[57] Guo S, Heinke R, Stöckel S, Rösch P, Popp J, Bocklitz T. Model transfer for Raman-spectroscopy-based bacterial classification. J Raman Spectrosc. 2018;49:627–37.
[58] Kalivas JH, Siano GG, Andries E, Goicoechea HC. Calibration maintenance and transfer using Tikhonov regularization approaches. Appl Spectrosc. 2009;63:800–9.
[59] Liang C, Yuan H-F, Zhao Z, Song C-F, Wang J-J. A new multivariate calibration model transfer method of near-infrared spectral analysis. Chemometrics Intell Lab Syst. 2016;153:51–7.
[60] Bloemberg TG, Gerretzen J, Lunshof A, Wehrens R, Buydens LM. Warping methods for spectroscopic and chromatographic signal alignment: a tutorial. Anal Chim Acta. 2013;781:14–32.
[61] Kalivas JH, Brownfield B, Karki BJ. Sample-wise spectral multivariate calibration desensitized to new artifacts relative to the calibration data using a residual penalty. J Chemom. 2017;31:e2873
[62] Bevilacqua M, Marini F. Local classification: locally weighted-partial least squares-discriminant analysis (LW-PLS-DA). Anal Chim Acta. 2014;838:20–30.

[63] Kalivas JH. Overview of two-norm (L2) and one-norm (L1) Tikhonov regularization variants for full wavelength or sparse spectral multivariate calibration models or maintenance. J Chemom. 2012;26:218–30.
[64] Castanedo F. A review of data fusion techniques. Sci World J. 2013;2013:19.
[65] Márquez C, López MI, Ruisánchez I, Callao MP. FT-Raman and NIR spectroscopy data fusion strategy for multivariate qualitative analysis of food fraud. Talanta. 2016;161:80–6.
[66] Teglia CM, Azcarate SM, Alcaráz MR, Goicoechea HC, Culzoni MJ. Exploiting the synergistic effect of concurrent data signals: low-level fusion of liquid chromatographic with dual detection data. Talanta. 2018;186:481–8.
[67] Borràs E, Ferré J, Boqué R, Mestres M, Aceña L, Busto O. Data fusion methodologies for food and beverage authentication and quality assessment – a review. Anal Chim Acta. 2015;891:1–14.
[68] Durrant-Whyte H, Stevens M, Nettleton E. Data fusion in decentralised sensing networks. In: 4th International Conference on Information Fusion, 2001.
[69] Bocklitz T, Bräutigam K, Urbanek A, Hoffmann F, Von Eggeling F, Ernst G, et al. Novel workflow for combining Raman spectroscopy and MALDI-MSI for tissue based studies. Anal Bioanal Chem. 2015;407:7865–73.
[70] Bocklitz T, Crecelius AC, Matthaus C, Tarcea N, Von Eggeling F, Schmitt M, et al. Deeper understanding of biological tissue: quantitative correlation of MALDI-TOF and Raman imaging. Anal Chem. 2013;85:10829–34.

Nathalie Jung and Maike Windbergs

5 Raman spectroscopy in pharmaceutical research and industry

Abstract: In the fast-developing fields of pharmaceutical research and industry, the implementation of Raman spectroscopy and related technologies has been very well received due to the combination of chemical selectivity and the option for non-invasive analysis of samples. This chapter explores established and potential applications of Raman spectroscopy, confocal Raman microscopy and related techniques from the early stages of drug development research up to the implementation of these techniques in process analytical technology (PAT) concepts for large-scale production in the pharmaceutical industry. Within this chapter, the implementation of Raman spectroscopy in the process of selection and optimisation of active pharmaceutical ingredients (APIs) and investigation of the interaction with excipients is described. Going beyond the scope of early drug development, the reader is introduced to the use of Raman techniques for the characterization of complex drug delivery systems, highlighting the technical requirements and describing the analysis of qualitative and quantitative composition as well as spatial component distribution within these pharmaceutical systems. Further, the reader is introduced to the application of Raman techniques for performance testing of drug delivery systems addressing drug release kinetics and interactions with biological systems ranging from single cells up to complex tissues. In the last part of this chapter, the advantages and recent developments of integrating Raman technologies into PAT processes for solid drug delivery systems and biologically derived pharmaceutics are discussed, demonstrating the impact of the technique on current quality control standards in industrial production and providing good prospects for future developments in the field of quality control at the terminal part of the supply chain and various other fields like individualized medicine.

On the way from the active drug molecule (API) in the research laboratory to the marketed medicine in the pharmacy, therapeutic efficacy of the active molecule and safety of the final medicine for the patient are of utmost importance. For each step, strict regulatory requirements apply which demand for suitable analytical techniques to acquire robust data to understand and control design, manufacturing and industrial large-scale production of medicines. In this context, Raman spectroscopy has come to the fore due to the combination of chemical selectivity and the option for non-invasive analysis of samples. Following the technical advancements in Raman equipment and

This article has previously been published in the journal *Physical Sciences Reviews*. Please cite as: Miliani, C., Rosi, F., Cartechini, L., Sali, D., Raman spectroscopy in pharmaceutical research and industry *Physical Sciences Reviews* [Online] **2019**, 4. DOI: 10.1515/psr-2017-0045.

https://doi.org/10.1515/9783110515312-005

analysis software, Raman spectroscopy and microscopy proofed to be valuable methods with versatile applications in pharmaceutical research and industry, starting from the analysis of single drug molecules as well as complex multi-component formulations up to automatized quality control during industrial production.

Keywords: Pharmaceutical research, Pharmaceutical idustry, Process analytical technology (PAT), Drug delivery

5.1 Raman spectroscopy in early drug development and preformulation

5.1.1 Identification of active pharmaceutical ingredients (APIs)

Despite continuous progress in pharmaceutical sciences and industry, global health care is confronted with serious challenges. The growing world population in combination with the increasing prevalence of severe diseases like cancer as well as the appearance of epidemics with resistant pathogens require the development of novel therapeutics. The process of identifying effective novel drugs and their molecular optimization is elaborate and is typically achieved through a combination of computational approaches and high throughput methodologies [1]. Newly discovered molecules that hold potential for therapeutic applications need to be thoroughly characterized with respect to their pharmacological effect as well as potential instabilities of the molecule before they can be incorporated into applicable formulations (like tablets or liquids for injection) to ensure their stability during fabrication and storage, and thus efficacy and safety for the patients´ therapy. In order to obtain a complete understanding of the molecular structure of novel active molecules and their potential modifications (induced by external factors like humidity, temperature etc.), the substance is commonly analysed using established spectroscopic methods like mass spectrometry (MS) and nuclear magnetic resonance spectroscopy (NMR). In this context, vibrational spectroscopic methods like Fourier-transformed infrared spectroscopy (FT-IR) and Raman spectroscopy can be valuable complementary tools. However, the interpretation of the resulting spectra is complex and requires expert knowledge. Therefore, Raman spectroscopy is mostly employed after the basic molecular structure has been assessed via MS and NMR to elucidate the absolute configuration of the molecule. So far, single-crystal X-ray diffraction (SCXRD) is known to be the gold standard for structural investigations, providing detailed information which are unparalleled. However, sample preparation involves crystallisation of the molecule, which is often not feasible. In this context, Raman spectroscopy serves as a beneficial complementary technique, which is able to identify different chiral centres of the molecule by using Raman optical activity (ROA) [2, 3]. Here, the molecule is irradiated with left and right polarised lasers and a difference spectrum is generated which can be compared to calculated spectra for each possible molecule conformation. Furthermore, the unique ROA band of a molecule can be used to identify its preferred enantiomeric form within a mixture of different forms [4], even in a biological relevant environment, since Raman spectroscopy is a non-invasive technique and does not require further sample preparation.

Due to the fact that Raman spectroscopy (in contrast to IR spectroscopy) is barely influenced by the presence of water, analysis in aqueous environment is possible, allowing for solubility experiments as well as interaction studies with biological systems like cells. Raman spectroscopy has already proven its suitability for studying proteins in solution as well as in crystalline solid state [5, 6]. As more and more biological macromolecules arise from the pipeline during the search for new APIs, comprehensive analysis of such molecules is of major importance. Using Raman spectroscopy, a multitude of conformational states and transformations on the protein level can be observed like denaturation [7–9], protein aggregation and secondary protein structures [10], as well as characterization of cysteine side chains [11]. Applying ROA, the stereochemistry of the amino acids constituting the protein can be monitored providing crucial information on the structure of the molecule. Hence, Raman spectroscopy has come to the fore for identification of active molecules produced in the biotechnology industry and during bioprocessing. Raman spectroscopy can be employed in a multitude of different processes starting with simple in-line concentration measurements of glucose and ethanol during the conversion through yeast as a fast alternative to the general time-consuming chromatography approach [12] up to the identification of biological pharmaceuticals like antibiotics [13] and quantification of posttranslational modification processes [14].

5.1.2 Selection and optimisation of APIs

The APIs that are selected and characterised in the discovery process are further studied during drug development determining the optimal form of the molecule to guarantee pharmacological efficiency as well as physical stability of the molecule in the final formulation. This process includes the assessment of all possible polymorphic forms (different crystal forms of the same molecule) of the API since the form has direct implications for the physicochemical properties of the molecule, resulting in differences in stability and solubility that could jeopardise the efficacy and safety of the final drug product. Drug polymorphism is often assessed via well established methods like differential scanning calorimetry (DSC), X-ray powder diffraction (XRPD) or NMR which provide valuable information, but require extensive sample preparation and time-consuming experimental procedures. In this context, Raman spectroscopy combines advantages in speed and ease of sample preparation with conclusive insight into the chemical structure of the different forms [15]. Raman spectroscopy even allows for real-time analysis of polymorph formation in microcrystals [16]. Since the formation of polymorphs and hydrates is very much dependent on physical and thermal conditions, it is of interest to monitor the relationship between the different forms of a drug and its behaviour during the manufacturing process and under storage conditions. Standard production procedures like tabletting and micronisation (involving physical stress) as well as contact with liquids during wet granulation for example might lead to the formation of unwanted API forms that may obstruct the pharmacological effect and stability of the final drug product. Besides NIR spectroscopy, Raman spectroscopy in combination with a heating stage can be used to monitor thermally induced conformational changes of the compound in real-time [17]. However, employing this setup the user needs to take into consideration that Raman spectra are sensitive to temperature changes and that peak position and shape can be influenced by the experimental conditions [18]. By expanding the instrumental setup with a humidity control chamber, the influence of relative humidity may also be taken into account during the experiment [19, 20]. Further, the influence of compression on polymorphic transitions of an API during a grinding process and hydrodynamic pressuring was assessed by the respective Raman spectra and a correlation of pressure and the corresponding changes in the Raman spectra could successfully be established as Figure 5.1 displays [19, 20].

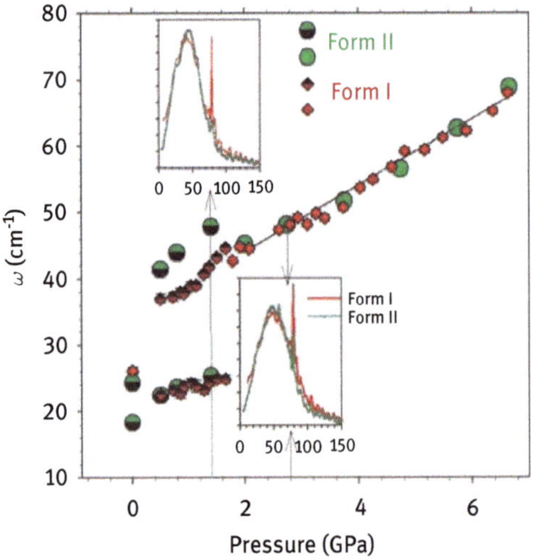

Figure 5.1: The correlation of pressure and changes in Raman spectra acquired from two different forms of anhydrous caffeine at room temperature.
Reprinted with permission from Elsevier [20].

Rather than modifying a microscope setup, a Raman probe can also be incorporated into a DSC or a dynamic vapour sorption (DVS) instrument to monitor conformational changes via the Raman signal of macrosamples [21, 22] even in real-time [23].

Structural changes of an API like the formation of salts are investigated during preformulation, since these alterations also influence the solubility and thereby finally also the bioavailability of the drug in the human body. Salification and co-crystallising of the drug are used to modify the solubility and stability of the molecule, but since the optimal counter ion cannot be determined via computational methods, elaborate salt screens are used, where a variety of different solvents is used to modify the final drug form to display desired properties [24]. The drug crystals are analysed via optical (polarized light microscopy) or thermal methods like DSC and thermogravimetric analysis (TGA) but the gold standard remains single crystal XRD which provides the most detailed structural information [25, 26]. However, this technique requires single drug crystals for the analysis, which often poses a problem for sample preparation. Raman spectroscopy on the other hand excels in the analysis of salt screens due to its sensitivity to molecular geometry and lack of sample preparation as well as very fast experimental procedures [27, 28]. With regard to the evolving technical advancements of Raman spectroscopy, an automated analysis provides valuable prerequisites for high-throughput analysis of API forms in microfluidic polymorph and co-crystal screenings [29, 30].

5.1.3 Interaction of APIs with excipients

As most drugs cannot be administered in their native (powder) form, drug delivery systems like tablets are developed to incorporate the required dose of the API, potentially also involving protection of the molecule against external influences (e. g. pH changes). A crucial factor during the development of novel therapeutics is the interaction of drugs with additives that form or are additionally incorporated into the final formulation. It must be taken into consideration that stabilising and solubilising agents, binder and penetration enhancer could possibly interact and change the structure of the drug molecule or even degrade it, thereby interfering with the efficacy of the drug product. During the extensive interaction studies, high performance liquid chromatography (HPLC) and liquid-chromatography mass spectrometry (LC-MS) are currently the methods of choice to elucidate the full extent of drug degradation. As part of ongoing progress in drug characterisation methodologies, spectroscopic methods were assessed for the characterisation of degradation processes, where Raman spectroscopy was found to be the most sensitive technique to detect molecular differences occurring during interactions [31]. Further, due to its non-invasive nature, Raman spectroscopy proves to be useful to realise the conduct of interaction and degradation studies in real-time, like the influence of pH changes to the kinetics of redox reactions of a drug compound via online Raman spectroscopy [32].

5.2 Characterization and evaluation of advanced drug delivery systems

5.2.1 Technical requirements for imaging pharmaceutical samples

For the in-depth analysis of advanced drug delivery systems incorporating the API, the acquisition of simple Raman spectra might not be sufficient as in many cases spatially resolved data are required. To characterize these systems thoroughly including composition and component distribution, physical stability and drug release from the carrier system up to interactions with biological systems derived from the human body like cells and tissues, advanced technical setups are required. The combination of Raman spectroscopy with a confocal microscope setup opened up a new horizon and paved the way for various pharmaceutical applications. However, many pharmaceutical samples exhibit a structured outer surface and imaging with a confocal microscope is limited to the respective focal plane. Thus, complementary techniques for determining the surface structure prior or simultaneously to the acquisition of Raman spectra were required. For this purpose, optical topography analysis based on white light reflection (Figure 5.2) and atomic force microscopy (AFM) were identified as suitable analytical approaches and technical advancements meanwhile allowed for combination of these techniques and Raman

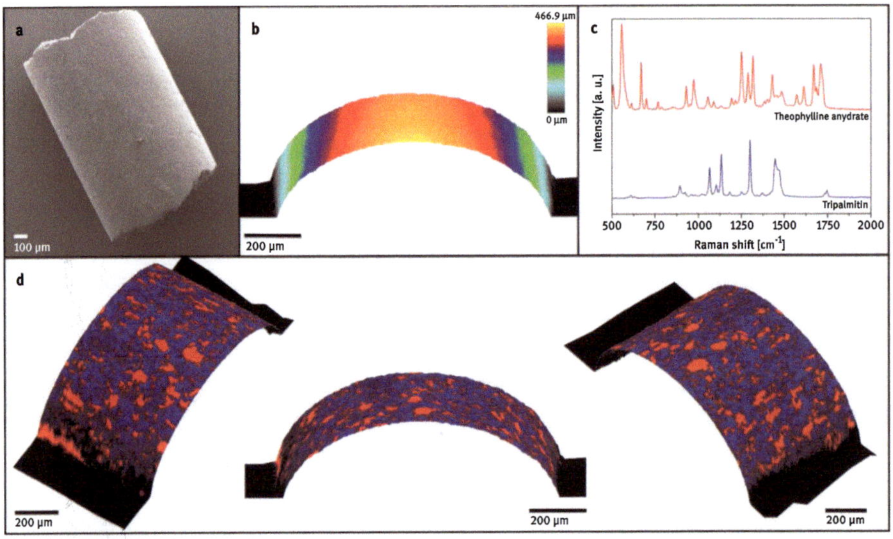

Figure 5.2: The combination of optical topography and Raman analysis of a drug-loaded extrudate. (a) Electron micrograph of the extrudate, (b) topography profile of the outer surface, (c) Raman spectra of API and excipient, (d) overlay of topography profile and Raman analysis displayed in false colours (red - API, blue – excipient).
Reprinted by permission from Springer Nature [35].

microscopy in one analytical platform [33–35]. Consequently, software capable of processing and combining such data is also required.

5.2.2 Composition and physicochemical stability of drug delivery systems

The determination of the chemical composition of a final drug product is a critical attribute for the quality of marketed therapeutics. During the fabrication process of pharmaceutical products, mechanical and thermal forces potentially lead to physicochemical changes of the active ingredient or excipients. Differences in drug content and polymorphic state of drugs might lead to ineffective treatment or might even be harmful to the patient. Therefore, a comprehensive examination of pharmaceuticals after manufacturing constitutes an inevitable step to ensure product safety.

The chemical identity and the distribution of a drug in low dosage pharmaceutical tablets were assessed in a study by Henson and Zhang [36]. They successfully detected two undesired polymorphic forms of the API in concentrations as low as 0.05 % (w/w) within tablets, highlighting Raman micro-spectroscopy mapping as a valuable technique for the detection of polymorphic impurities. A similar study was conducted within lipid based inhalable powder formulations where content uniformity and the polymorphic form of a drug substance were determined [37]. In a study on two different solid dispersions, Raman spectroscopy was used to evaluate drug stability over time and it was successfully shown that recrystallization occurred in one of the formulations, while for the comparator formulation such events could not be detected [38].

Further, Raman spectroscopy was combined with SEM and TEM to evaluate if a poorly water-soluble drug is present as an amorphous nanodispersion or if it exists as a molecularly dispersed compound within a polymeric matrix [39]. Raman mapping was able to elucidate the size and spatial distribution of areas where the drug existed in its molecularly dispersed form or in nano crystals, respectively. Sievens-Figueroa et al. characterized edible polymer films containing drug nanoparticles and used Raman microscopy complementary to XRPD, showing that the film processing methodology had no negative effect on the drug crystallinity [40]. The assessment of the chemical identity as well as particle size and particle size distribution were subject of the investigations performed in another study on aqueous nasal spray suspension formulations by Doub et al. [41]. In a study on two verapamil formulations, Raman microscopy was used to investigate composition and distribution of drug and excipients, and it was found that the composition of the formulations was identical and differences in release kinetics were solely due to the modifications in the manufacturing process [42].

Due to its chemical selectivity, Raman spectroscopy proves to be an invaluable tool for the assessment of compositional identity and, therefore, is experiencing an incline in popularity in the identification and analysis of counterfeit products.

Rebiere et al. developed a method for Raman chemical imaging of counterfeit formulations, which enabled differentiation of three polymorphs of clopidogrel as well as direct quantification of API content [43]. Along the same lines, another study described the development of a method for the detection of sildenafil in tablets. The authors used a handheld Raman spectrometer to measure sample solutions after dissolution directly though glass vials without the need for further sample preparation [44]. The detection of counterfeits is not limited to delivery systems incorporating small molecules, but is also emerging as a valuable tool for the identification of falsified protein-based medicines. Raman spectroscopy and microscopy were used in conjunction with optimized analytical methods to evaluate 12 substances in liquid or lyophilized form directly through glass packaging and compared them to their counterfeits in a qualitative and quantitative manner [45].

5.2.3 Spatial component distribution within drug delivery systems

While the chemical composition of pharmaceuticals can also be determined via destructive methods like HPLC, the non-invasive Raman approach stands out when it comes to analysis of the spatial distribution of drug and excipients within a drug product. Advancements in instrumentation, computer technology and data analysis allowed for the development of other Raman spectroscopy based techniques like coherent anti-Stokes Raman spectroscopy (CARS), stimulated Raman spectroscopy (SRS) and surface enhanced Raman spectroscopy (SERS) and strengthened the interest in Raman microscopy as an analytical tool for pharmaceutical content analysis [46]. Unlike simpler content analyses with Raman spectroscopy, the chemically-selective determination of the spatial compound distribution requires suitable technical prerequisites for high resolution confocal Raman imaging.

Breitenbach et al. already investigated the spatial distribution of components in the 1990s, analysing a solid dispersion of ibuprofen in polyvinylpyrrolidone and showed that Raman microscopy is suited to verify homogeneous drug distribution in the system [47]. Raman microscopy was further applied to biodegradable double-walled microspheres, for which the investigators analysed cross sections of the samples and determined chemical composition and spatial distribution of the different compounds [48]. A similar approach was used to characterize hormone loaded sucrose spheres [49]. Raman mapping was used to show the integrity of the enteric coating and assess the distribution of a hormone drug beneath the coating within the particles. Šašić et al. applied Raman spectroscopy on a larger scale by examining pharmaceutical tablets that varied in the manufacturing process and analysed the distribution of all ingredients throughout the tablet, showing the superiority of Raman microscopy over NIR imaging [50]. Along the same lines, another study used optical topography to map the surface of 3D printed tablets and further analyse the sample which was loaded with nanocapsules using confocal Raman microscopy [51].

The technique was able to analyse the spatial distribution of mannitol and microcrystalline cellulose in the tablet during crucial manufacturing steps. Surface topography measurements with AFM were also combined with confocal Raman microscopy to study the distributions of plasticizer and drug within topical polymeric film-forming systems [52]. In another correlative approach, confocal Raman microscopy was used in conjunction with AFM to connect surface topography with chemical imaging in the characterization of a drug-eluting stent. The study showed for the first time a correlation between coating morphology and drug concentration in the medical device [53]. Raman imaging was further applied to characterize the distribution of single components in semi-solid self-emulsifying systems for oral drug delivery and to evaluate the homogeneity of the dosage form [54].

5.2.4 Drug release from delivery systems

Drug release from dosage forms and their kinetics are crucial characteristics which are of major importance during the development of novel therapeutics, as the time-dependant availability of the drug in the human body is a prerequisite for effective therapy. Based on the monographs in the Pharmacopeia, drug release is assessed by quantitative methods like HPLC or UV/VIS spectroscopy to gain quantitative drug release profiles over time. As Raman spectroscopy is a linear technique, the acquisition of Raman spectra over time results in similar profiles. However, even more important, Raman spectroscopy/microscopy is capable of monitoring changes of the drug upon contact with the release medium over time (e. g. hydration) as well as changes in component distribution within the drug delivery system and spatially resolved drug depletion. In a recent study, the release mechanism of a poorly water-soluble drug from different solid dispersions was investigated using confocal Raman microscopy [55]. The technique helped to elucidate the influence of the utilized polymer on the release kinetics of the drug. In a similar study, the distribution of ibuprofen and its release were imaged in different types of microparticles [56]. The examination of the samples using Raman microscopy enabled the investigators to improve their formulation to obtain desired release properties. As shown in a study by Vukosavljević et al., confocal Raman microscopy proves to be a useful tool for the monitoring of dissolution processes of coated pellets [57]. Figure 5.3 depicts virtual cross sections of the pellets before and after release testing and visualises the mechanism of pore formation that enables the liberation of the drug molecule from the dosage form.

In another study, CARS was also found to be very useful for real-time monitoring the transformation of theophylline anhydrate to monohydrate on the surface of tablets during dissolution testing [58]. Further, drug release from a complex geometrical three-layered tablet was investigated in real-time using Raman microscopy to elucidate how the drug is being released in a sustained manner [59]. Drug release

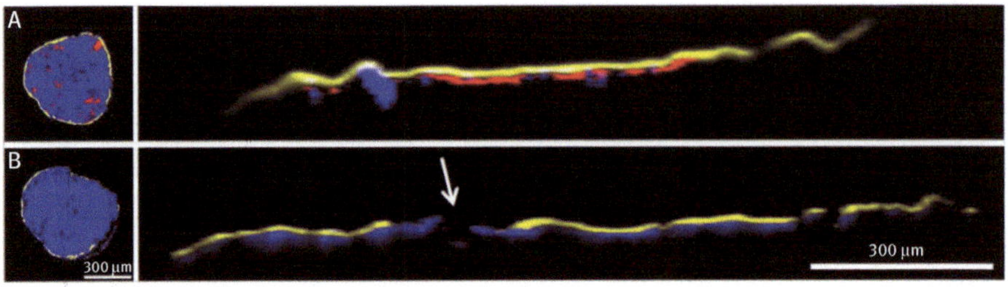

Figure 5.3: Raman mapping of cross sections of coated pellets before (A) and after (B) in vitro release testing. Film coating is depicted in yellow, matrix excipients in blue and drug in red, respectively. Reprinted with permission from John Wiley and Sons [57].

and morphological changes of a stent coating upon elution was also investigated using confocal Raman microscopy in conjunction with other methods like in-liquid AFM [53, 60]. The release of drug could be correlated with morphological changes of the polymer matrix and the formation of pore networks.

5.2.5 Interaction with biological systems

In addition to thorough analysis of all physicochemical characteristics and the drug release mechanism as well as the corresponding release kinetics, interactions of the therapeutic with biological systems have to be investigated simulating the application route as well as the pharmacological target in the human body (if possible). For this purpose, artificial systems as well as cell- and tissue-based assays are applied. In contrast to fluorescence microscopy which requires the implementation of bulky fluorescent labels potentially altering the physiological reaction of the drug with cells or tissues, Raman microscopy offers the remarkable advantage of chemically-selective imaging without disturbance of natural processes or alteration of the biological system [61]. In this context, the uptake of liposomal drug carrier systems into cells was studies in two case studies where Raman microscopy was able to follow the internalization of liposomes into cells via the usage of a deuterium tag [62] and even visualize specified mitochondrial targeting by chemically modifying the liposomes [63]. With respect to rising interest in the relevance of nanoparticles in health issues and their utilization as pharmaceutical drug carrier systems, the uptake of titanium dioxide in oral buccal epithelia cells as well as in lung cells was investigated by two different groups, showing that the technique enables the localization of nanoparticles within human cells [64, 65]. As shown in Figure 5.4, titanium dioxide particles could be visualized in virtual cross sections of the cells in the cytosol, as well as within the cell nucleus.

Further studies on nanoparticle uptake by cells were performed using silver nanoparticles and following their interaction with human fibroblasts [66]. Raman

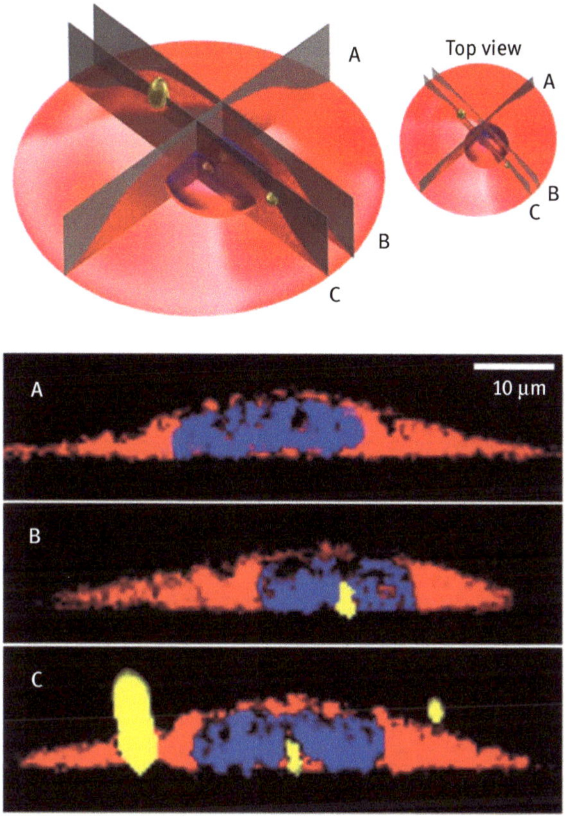

Figure 5.4: Schematic sketch of an oral buccal epithelia cell and virtual cross sections through the cell acquired by confocal Raman imaging. Cell bodies are depicted in red, nuclei in blue and titandioxide particles in yellow. (A) Cross section of cell without particles. (B) Titanium dioxide particle within nucleus. (C) Multiple titanium dioxide particles outside and within the cell. Reprinted from [65] with permission of The Royal Society of Chemistry.

microscopy was used to follow penetration of the particles into skin cells and localizing them within the cytosol. The visualization of nanoparticles within biological tissues poses another problem due to the complex composition and optical properties of tissues in contrast to single cells. This issue was overcome in a study using deuterated chitosan nanoparticles and CARS microscopy to detect the uptake into brain and liver tissue of mice [67]. The incorporation of Raman active compounds into liposomes and nanoparticles resembles an elegant mechanism for tracking and visualization of these drug carriers. However, when it comes to the detection of protein compounds, other solutions for the distinction from the biological system are required. In a study by Ye et al. C^{13} was used to modify phenylalanine in the protein penetratin to study its uptake and distribution in melanoma cells [68].

Recently confocal Raman microscopy was implemented in the analysis of semi-solid topical formulations and their penetration into skin tissue. Franzen et al. published a series of comprehensive studies, analysing the Raman spectral variability of human *stratum corneum* and explaining the methodology of quantification of a model drug in excised human skin as well as evaluating skin derived Raman peaks as internal standards for peak correlations [69, 70]. The penetration abilities of caffeine nanocrystals and propylene glycol as well as a possible synergetic effect was assessed using confocal Raman microscopy on excised porcine skin by measuring penetration depths of both substances [71]. Zhang et al. further refined the analysis of penetration and studied the interaction of the drug 5-fluorouracil with excised skin tissue and observed the interaction of the substance with the biological tissue [72]. Further, as shown in Figure 5.5, confocal Raman microscopy was able to detect

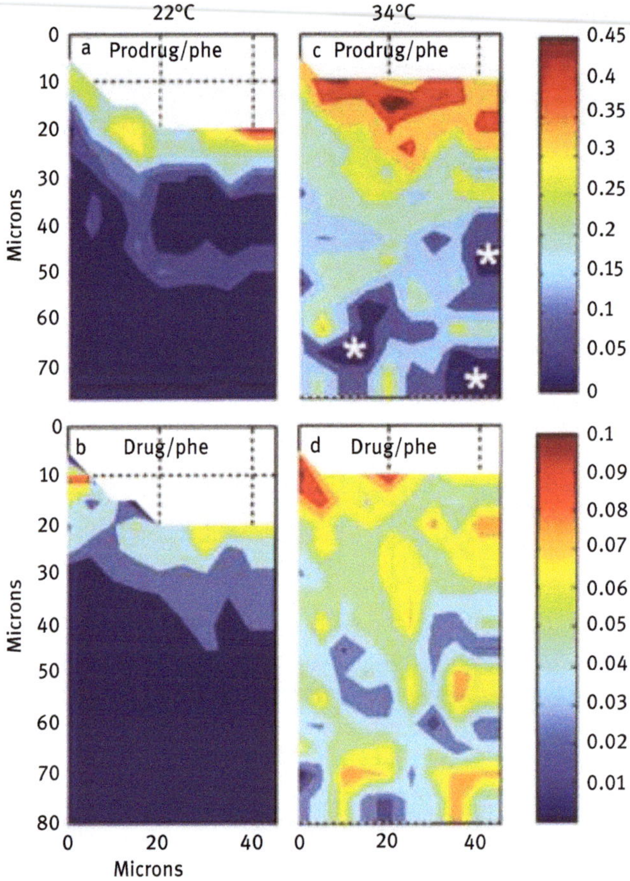

Figure 5.5: Virtual cross sections of pig skin showing the distribution of prodrug and active drug after incubation of the samples at 22 °C (a and c) and 34 °C (b and d).
Reprinted in part with permission from Elsevier [72].

the conversion of the prodrug into the active form and visualize the distribution of both compounds in virtual cross sections.

Further studies on the penetration enhancing effect of different substances on flufenamic acid were accomplished in excised human skin and delivered valuable insights into drug penetration kinetics into and through the human skin barrier [73]. Similar studies were also undertaken *in vivo*, investigating the effect of penetration enhancer on retinol delivery through human *stratum corneum*, highlighting once again the advantage of non-invasive and label-free imaging capabilities of Raman microscopy [74].

5.3 Process analytical technology (PAT) and quality control during pharmaceutical manufacturing

The term "Process Analytical Technology" was introduced in the context of a general concept to design, analyse and finally control the manufacturing of pharmaceuticals. For this approach, critical quality attributes for a distinct pharmaceutical are defined and have to be controlled by the monitoring of critical process parameters affecting the critical quality attributes of the respective pharmaceutical product during manufacturing. To realize this concept, suitable analytical techniques are required, preferentially allowing for continuous on-line or in-line measurements. However, the in-line analysis of fabrication processes in real time poses a challenge for traditional analytical methods as HPLC and similar techniques that require time- and work-intensive sample preparation. Based on its characteristics like chemical selectivity as well as the non-invasive nature of data acquisition, Raman spectroscopy is well suited for PAT applications, as recent studies report [75]. With its diversified application area for the analysis of solids, semi-solids and even biologically derived pharmaceuticals and its fast data acquisition, Raman spectroscopy has gained the interest of the industry to improve the efficiency of manufacturing processes.

5.3.1 PAT for the manufacturing of solid drug delivery systems

The crystallization of an active compound is a crucial step in the manufacturing of pharmaceuticals. Monitoring of crystallization processes of drug molecules and identification of different polymorphic forms which might eventually occur during these processes is essential to ensure the generation of the pharmacologically effective polymorphic form and thereby assure therapeutic efficiency. Raman spectroscopy with its high chemical selectivity holds great potential for the determination of different polymorphic forms and is, therefore, successfully being applied during crystallization processes in pharmaceutical sciences and industry. By applying a good calibration practice procedure, described in a study by Simone et al., the quantitative assessment of polymorphic transformation processes within solutions can be achieved by using Raman spectroscopy [76]. The implementation of in-line Raman spectroscopy was described in a study, where cooling crystallization of a drug was performed [77]. Raman enabled the real-time monitoring of the crystallization process and gave continuous feedback on the polymorphic forms present in the sample, ensuring maintenance of drug quality during the process.

The manufacturing of tablets as well as of capsules involves several steps holding the potential to alter and modify the starting material in undesired ways. In-line monitoring and evaluation of each intermediate (e. g. during homogenization and blending processes) leading to the final product presents a necessary step towards conscientious manufacturing of pharmaceuticals. In this context, the implementation of a Raman fibre probe for in-line monitoring of the homogeneity of an aqueous pharmaceutical suspension was described by De Beer et al. [78]. The technique was not only able to monitor the homogenization, but also helped to understand and control the process and further allowed for the real-time quantification of the drug during mixing of the suspension. The successful implementation of Raman spectroscopy in solid dosage form manufacturing was shown in studies by De Beer et al., who studied a powder blending process with a fibre optical immersion Raman probe, taking into account process and formulation variables [79]. Raman spectroscopy helped to understand the blending process and was successfully implemented for the end point monitoring as confirmed by NIR spectroscopy as a second independent analytical method. Further, in-line Raman spectroscopy was used to monitor the blending process of a pharmaceutical three-component model system with respect to drug content and blend homogeneity [80]. Uniformity of the blend was successfully monitored and the Raman spectroscopy-based feedback control of the system allowed for automated dosing of the drug to reach the desired concentration of the formulation. Regarding these applications of Raman spectroscopy, the method presents an interesting alternative to the traditional HPLC analysis for assessing product homogeneity. A study by Riolo et al. showed that data acquired with Raman spectroscopy were superior to HPLC data in terms of data robustness [81].

Further, granulation processes hold potential for the application of Raman spectroscopy as well. *In-situ* real-time Raman spectroscopy was employed in one study to monitor a fluidised bed granulation process and to characterize the composition of the sample in three spatial dimensions as a function of time [82]. Another study describes the implementation of Raman spectroscopy during high-shear wet granulation to monitor the conversion of anhydrous theophylline to theophylline monohydrate and shows that the technique is a powerful tool to observe solid-state transformation kinetics during the manufacturing process as shown in Figure 5.6 [83].

The monitoring of potential solid state changes is further necessary after and during drying processes. Raman spectroscopy was assessed together with NIR spectroscopy as an alternative to the traditional measurement of product temperature or the humidity of the outlet drying air during fluid bed drying [84]. The combination

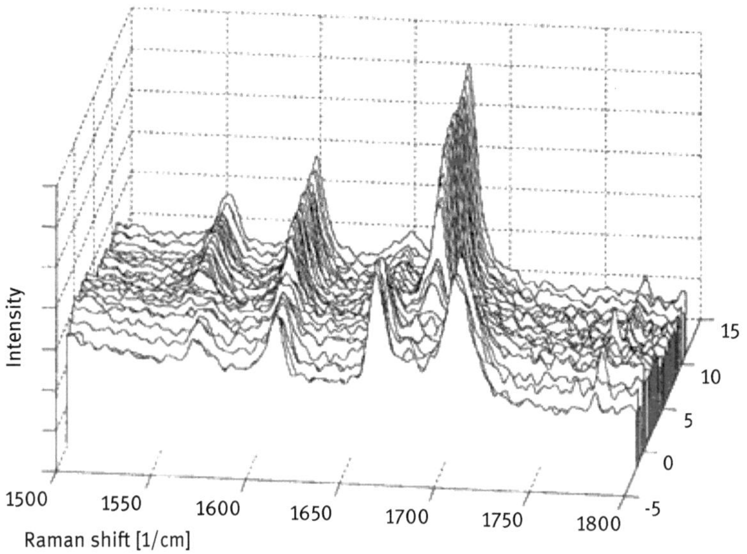

Figure 5.6: Raman spectra displaying the transformation of anhydrous theophylline (characteristic peaks at 1664 and 1707 cm^{-1}) to theophylline monohydrate (1686 cm^{-1}) during a wet granulation process.
Reprinted from [83] with permission from Elsevier.

of both methods was found to be able to determine moisture content as well as chemically characterize the sample in-line and thereby represents a valuable alternative to the traditional methods. Kogermann et al. further refined this approach by carrying out a quantitative solid phase analysis of two drug compounds within their sample using Raman spectroscopy and partial least squares (PLS) regression [85]. TGA and XRPD were used to confirm the quantification approach via Raman spectroscopy and the results were found to be in reasonable agreement with each other, showing that Raman might also be implemented for quantitative in-line measurements during drying processes in industrial setups.

The successful application of in-line Raman spectroscopy for monitoring freeze drying was presented in a study by De Beer et al., where Raman and NIR probes were implemented in a freeze drying chamber [86]. The combination of both methods allowed for a very comprehensive evaluation of the freeze drying process by analysing the solid state of a mannitol solution as well as the onset of ice nucleation and drug crystallization as well as the endpoint of ice sublimation.

The coating process of a solid oral dosage form requires the assessment of coating thickness and integrity to ensure a flawless final drug product. In order to streamline and improve the workflow and quality assessment in industrial manufacturing, as alternative methods to the traditional visual inspection and weight measurements, Raman spectroscopy and other spectroscopic methods were evaluated to investigate the coating procedure of tablets and pellets [87]. Especially the process of active coating requires strict monitoring in order to ensure an efficient incorporation of the active ingredient in the coating layer and a consistent covering of the core of the tablet. The in-line monitoring of an active coating process via Raman spectroscopy was described in a study where the spectral data were successfully correlated to the average weight gain and the amount of coated diprophylline at different measuring time points [88].

Due to rising popularity of hot melt extrusion for the fabrication of solid dispersions, the interest in implementing PAT technologies and thereby Raman spectroscopy into the manufacturing process has been investigated in multiple case studies. Saerens et al. investigated the incorporation of a Raman probe into a die head of a development-scaled twin-screw extruder for in-line solid state prediction, monitoring the influence of drug concentration, screw configurations, barrel temperature and pressure [89]. Raman spectroscopy was able to differentiate between glassy solid solutions and crystalline dispersions with a higher sensitivity than off-line XRPD and DSC analysis. The approach was developed further into a quantification measurement where a PLS model, regressing the drug concentration versus the in-line acquired Raman spectra, was validated [90]. Additionally, the transformation of the drug component and its interactions with the polymer were assessed during the hot melt extrusion process.

5.3.2 PAT for biologically-derived pharmaceuticals

Biologically-derived macromolecular drugs are often produced in fermentation processes or bioprocesses based on mammalian cells. The monitoring of drug production that yield rather low concentrations, such as the synthesis of antibiotics, require advanced analytics to improve the in-line analysis on an industrial scale manufacturing process. The detection of small molecules in aqueous environments requires an increase in sensitivity for a common Raman spectroscope. A sensitive *in-situ* detection of polar antibiotics was achieved by using surface-enhanced Raman spectroscopy (SERS) in combination with electrophoretic preconcentration, which allowed for the detection of different polar antibiotics in sub-nM concentrations [91]. Another study describes the quantification of five different penicillins within a fermentation broth using Raman spectroscopy [13]. While standard Raman spectroscopy allowed for the distinction between the different forms of the antibiotic and their quantification only within higher concentrations (>50 mM), SERS in combination with silver colloids presented an improved signal quality from considerably lower concentrations of penicillin and thereby highlights the method for the quantification of secondary metabolites in microbial processes. Figure 5.7 shows the difference between Raman spectroscopy and SERS during the analysis of penicillins.

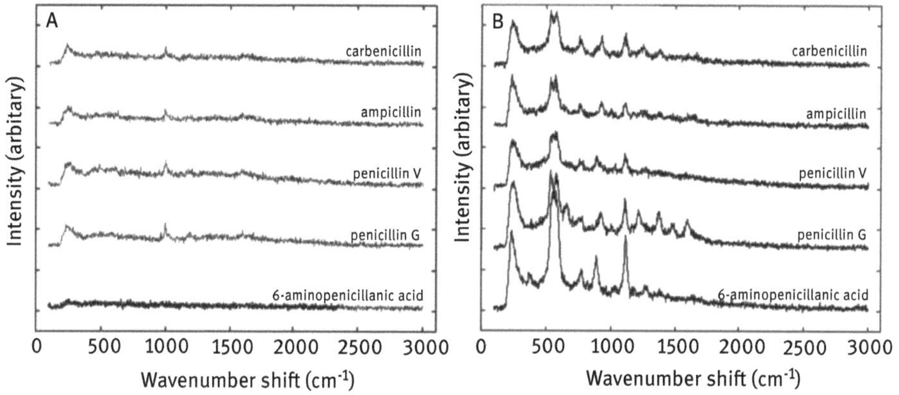

Figure 5.7: Difference between Raman (A) and SERS (B) spectra of the five different penicillin forms (62.5 mM). Spectra were accumulated with 10 sec integration time and 1 accumulation each. Adapted from [13] with permission of The Royal Society of Chemistry.

Since the use of biological material for the production of pharmaceuticals poses a special issue concerning contamination of growth media, the evaluation of the aqueous production environment needs to be closely monitored. Further, PAT technologies find application in the monitoring of cell culture media that present a crucial prerequisite for cell function and production of bio-macromolecules. The state and possible degradation of culture medium is a quality factor that needs proper monitoring in order to ensure a smooth manufacturing process and requires highly sensitive, in-line applicable technologies. In an attempt to meet these demands, SERS was evaluated as a method to monitor stability of cell culture media [92]. The generated chemometric data enabled the detection of compositional changes contingent on photodegradation in the media, once stringent reproducibility controls were implemented. Implementation of Raman spectroscopy to monitor multiple parameters of mammalian cell culture bioreactor processes were described in a study by Abu-Absi et al. [93]. A Raman probe was used to monitor compositional parameters like content of glutamine, glutamate, glucose, lactate, and ammonium, as well as cell densities. In-line measurement and analysis of different culture media parameters has further been used in a fed-batch fermentation procedure to predict the final product yield during all stages in an industrial setup [94]. Several small-scaled batches were analysed regarding glycoprotein yield to manage the selection of suitable batches for large-scale manufacture in order to improve the process efficiency. A similar approach was used in a different study to quantitatively analyse the production of active substances from biological manufacturing processes [95]. The in-line and real-time prediction of a recombinant antibody titer using a Raman immersion probe allowed for a robust measurement of the active substance while at the same time reducing the risk of contamination drastically compared to traditional sampling methods.

Biologically-derived pharmaceuticals constitute a particularly difficult challenge when it comes to proper storage of the produced macromolecules. Freeze drying has been found to be a suitable approach for long-term storage, whilst maintaining the structure and functionality of the molecules. The option of implementing Raman spectroscopy for the in-line and real-time quality assessment of biologically-derived macromolecules was considered in a study by Pieters et al., who used the technique to investigate the freeze drying process of lactate dehydrogenase in a lab-scale experiment [96]. They reported that Raman spectroscopy was able to detect changes in secondary protein structure and distinguish between native and non-native freeze-dried protein with 95 % accuracy, opening possibilities for the employment of Raman spectroscopy for automated analyses of this quality attribute for industrial processes.

5.4 Future perspectives of Raman spectroscopy in pharmaceutical research and industry

Due to technical advancements in spectroscopy setups as well as progress in multivariate-data analysis software, Raman techniques have experienced a drastic increase in interest for a wide variety of pharmaceutical applications. After initial implementation in research and development facilities, Raman based methods have paved their way into large scale pharmaceutical manufacturing, following the demands of regulatory frameworks and the PAT initiative. Before, incorporation of Raman spectrometers into manufacturing processes resembled an expensive and work-intensive approach, requiring specialists to analyse the acquired data sets. But with Raman probes becoming increasingly smaller and more advanced, their incorporation in single process vessels might fast become a standard in a variety of PAT related equipment, making Raman spectroscopy a widespread analysis technique in pharmaceutical manufacturing. The incorporation of the technique in high-end equipment and the integration of Raman derivatives like SRS, CARS and TERS will contribute to the advancement of industrial pharmaceutical development and production sites. Offering the possibility of in-line and real-time measurements, while simultaneously decreasing the risk of contamination and increasing the gain of production, the technique becomes more attractive to manufacturers with respect to return of their investment. Following the increasing demand for inbuilt Raman probes in various process steps, analysis software is evolving to accomplish automated measurements, paving the way to continuous pharmaceutical production within the quality by design concept. Establishment and access to Raman spectral libraries and advanced algorithms that will enable a conclusive automated interpretation of large Raman data sets may especially be important for the field of manufacturing of personalized and individualized drug delivery systems, where the need for an intelligent PAT tool may be most prevalent.

The implementation of Raman spectroscopic methods will further expand beyond the manufacturing process and will be integrated more efficiently for material qualification and quality control of pharmaceutical products at different stations of the supply chain. Incidents where falsified or modified drug products have been distributed by pharmacies, drew attention to the need for quality control at the end of the supply chain. Portable, low-resolution Raman instruments offer the possibility of monitoring the quality of tablets, semisolids as well as aqueous solutions and determining drug content as well as assessing degradation of products non-invasively through their packaging. The development of simplified analysis algorithms allows for the implementation of these systems even by untrained personnel and non-experts. Such devices present valuable assets, especially in regions with lower quality standards and the tendency towards distribution of counterfeit products.

The technical development will further progress into the high-end imaging equipment and evolve the current approaches of combining Raman microscopy with advanced imaging techniques like AFM and scanning near field microscopy as well as scanning electron microscopy [97]. The integration of Raman spectroscopy with these high-resolution approaches will enable a more detailed investigation of drug delivery systems and drug delivery processes in biological systems like tissue or even on the single cell level.

References

[1] Lipinski CA, Lombardo F, Dominy BW, Feeney PJ. Experimental and computational approaches to estimate solubility and permeability in drug discovery and development settings. Adv Drug Deliv Rev. 2001;46:3–26.

[2] Nafie LA. Infrared and Raman vibrational optical activity: theoretical and experimental aspects. Annu Rev Phys Chem. 1997;48:357–86.

[3] Qu X, Lee E, Yu G-S, Freedman TB, Nafie LA. Quantitative comparison of experimental infrared and Raman optical activity spectra. Appl Spectrosc. 1996;50:649–57.

[4] Spencer KM, Edmonds RB, Rauh RD. Analytical chiral purity verification using Raman optical activity. Appl Spectrosc. 1996;50:681–5.

[5] Tuma R. Raman spectroscopy of proteins: from peptides to large assemblies. Journal of Raman Spectroscopy. 2005;36:307–19.

[6] Sane SU, Wong R, Hsu CC. Raman spectroscopic characterization of drying-induced structural changes in a therapeutic antibody: correlating structural changes with long-term stability. J Pharm Sci. 2004;93:1005–18.

[7] Chi Z, Asher SA. UV resonance Raman determination of protein acid denaturation: selective unfolding of helical segments of horse myoglobin. Biochem. 1998;37:2865–72.

[8] Chi Z, Asher SA. Ultraviolet resonance Raman examination of horse apomyoglobin acid unfolding intermediates. Biochem. 1999;38:8196–203.

[9] Hédoux A, Paccou L, Guinet Y. Relationship between β-relaxation and structural stability of lysozyme: microscopic insight on thermostabilization mechanism by trehalose from Raman spectroscopy experiments. J Chem Phys. 2014;140:225102.

[10] Zheng R., Zheng X, Dong J, Carey PR. Proteins can convert to β-sheet in single crystals. Protein Sci. 2004;13:1288–94.
[11] Wen Z-Q, Cao X, Vance A. Conformation and side chains environments of recombinant human interleukin-1 receptor antagonist (rh-IL-1ra) probed by Raman, Raman optical activity, and UV-resonance Raman spectroscopy. J Pharm Sci. 2008;97:2228–41.
[12] Shaw AD, Kaderbhai N, Jones A, Woodward AM, Goodacre R, Rowland JJ, Kell DB. Noninvasive, on-line monitoring of the biotransformation by yeast of glucose to ethanol using dispersive Raman spectroscopy and chemometrics. Appl Spectrosc. 1999;53:1419–28.
[13] Clarke SJ, Littleford RE, Smith WE, Goodacre R. Rapid monitoring of antibiotics using Raman and surface enhanced Raman spectroscopy. Analyst. 2005;130:1019–26.
[14] Jarvis R.M., Blanch EW, Golovanov AP, Screen J, Goodacre R. Quantification of casein phosphorylation with conformational interpretation using Raman spectroscopy. Analyst. 2007;132:1053–60.
[15] Dandeu A., Humbert B, Carteret C, Muhr H, Plasari E, Bossoutrot JM. Raman spectroscopy – A powerful tool for the quantitative determination of the composition of polymorph mixtures: application to CaCO3 polymorph mixtures. Chem Eng Technol. 2006;29:221–5.
[16] Anquetil PA, Brenan CJ, Marcolli C, Hunter IW. Laser Raman spectroscopic analysis of polymorphic forms in microliter fluid volumes. J Pharm Sci. 2003;92:149–60.
[17] Szelagiewicz M, Marcolli C, Cianferani S, Hard A, Vit A, Burkhard A, von Raumer M, Hofmeier U, Zilian A, Francotte E, Schenker R. In situ Characterization of polymorphic forms: the potential of Raman techniques. J Therm Anal Calorim. 1999;57:23–43.
[18] Mehrens SM, Kale UJ, Qu X. Statistical analysis of differences in the Raman spectra of polymorphs. J Pharm Sci. 2005;94:1354–67.
[19] Clarke D, Physik A. Applications of vibrational spectroscopy in pharmaceutical research and development. In: Don JMC, Pivonka E, Griffiths PR, Editor. Applications of vibrational spectroscopy in pharmaceutical research and development. Chichester: Wiley, 2007.
[20] Hédoux A, Guinet Y, Paccou L, Danède F, Derollez P. Polymorphic transformation of anhydrous caffeine upon grinding and hydrostatic pressurizing analyzed by low-frequency Raman spectroscopy. J Pharm Sci. 2013;102:162–70.
[21] Kauffman JF, Batykefer LM, Tuschel DD. Raman detected differential scanning calorimetry of polymorphic transformations in acetaminophen. J Pharm Biomed Anal. 2008;48:1310–5.
[22] Raijada D., Bond AD, Larsen FH, Cornett C, Qu H, Rantanen J. Exploring the solid-form landscape of pharmaceutical hydrates: transformation pathways of the sodium naproxen anhydrate-hydrate system. Pharm Res. 2013;30:280–9.
[23] Wikström H, Rantanen J, Gift AD, Taylor LS. Toward an understanding of the factors influencing anhydrate-to-hydrate transformation kinetics in aqueous environments. Cryst Growth Des. 2008;8:2684–93.
[24] Newman A. Specialized solid form screening techniques. Org Process Res Dev. 2013;17:457–71.
[25] Datta S, Grant DJ. Crystal structures of drugs: advances in determination, prediction and engineering. Nat Rev Drug Discov. 2004;3:42–57.
[26] Stout GH, Jensen LH. X-ray structure determination, a practical guide. 2nd ed. New York: Chichester John Wiley & Sons, 1989.
[27] Kojima T, Tsutsumi S, Yamamoto K, Ikeda Y, Moriwaki T. High-throughput cocrystal slurry screening by use of in situ Raman microscopy and multi-well plate. Int J Pharm. 2010;399:52–9.
[28] Cerreia Vioglio P, Chierotti MR, Gobetto R. Pharmaceutical aspects of salt and cocrystal forms of APIs and characterization challenges. Adv Drug Deliv Rev. 2017;117:86–110.

[29] Remenar JF, MacPhee JM, Larson BK, Tyagi VA, Ho JH, McIlroy DA, Hickey MB, Shaw PB, Almarsson Ö. Salt selection and simultaneous polymorphism assessment via high-throughput crystallization: the case of sertraline. Org Process Res Dev. 2003;7:990–6.
[30] Peterson ML, Morissette SL, McNulty C, Goldsweig A, Shaw P, LeQuesne M, Monagle J, Encina N, Marchionna J, Johnson A, Gonzalez-Zugasti J, Lemmo AV, Ellis SJ, Cima MJ, Almarsson Ö. Iterative high-throughput polymorphism studies on acetaminophen and an experimentally derived structure for form III. J Am Chem Soc. 2002;124:10958–9.
[31] Feth MP, Nagel N, Baumgartner B, Brockelmann M, Rigal D, Otto B, Spitzenberg M, Schulz M, Becker B, Fischer F, Petzoldt C. Challenges in the development of hydrate phases as active pharmaceutical ingredients–an example. Eur J Pharm Sci. 2011;42:116–29.
[32] Skrdla PJ, Zhang D. Disproportionation of a crystalline citrate salt of a developmental pharmaceutical compound: characterization of the kinetics using pH monitoring and online Raman spectroscopy plus quantitation of the crystalline free base form in binary physical mixtures using FT-Raman, XRPD and DSC. J Pharm Biomed Anal. 2014;90:186–91.
[33] Prikulis J, Murty KV, Olin H, Kall M. Large-area topography analysis and near-field Raman spectroscopy using bent fibre probes. J Microsc. 2003;210:269–73.
[34] Robinson JA, Puls CP, Stanley NE, Stitt JP, Fanton MA. Raman topography and strain uniformity of large-area epitaxial graphene. Nano Lett. 2009;9:964–8.
[35] Kann B, Windbergs M. Chemical imaging of drug delivery systems with structured surfaces-a combined analytical approach of confocal raman microscopy and optical profilometry. AAPS J. 2013;15:1550–7416.
[36] Mark JH, Lin Z. Drug characterization in low dosage pharmaceutical tablets using Raman microscopic mapping. Appl Spectrosc. 2006;60:1247–55.
[37] Schoenherr C, Haefele T, Paulus K, Francese G. Confocal Raman microscopy to probe content uniformity of a lipid based powder for inhalation: A quality by design approach. Eur J Pharm Sci. 2009;38:47–54.
[38] Qian F, Huang J, Zhu Q, Haddadin R, Gawel J, Garmise R, Hussain M. Is a distinctive single Tg a reliable indicator for the homogeneity of amorphous solid dispersion? Int J Pharm. 2010;395:232–5.
[39] Karavas E, Georgarakis M, Docoslis A, Bikiaris D. Combining SEM, TEM, and micro-Raman techniques to differentiate between the amorphous molecular level dispersions and nanodispersions of a poorly water-soluble drug within a polymer matrix. Int J Pharm. 2007;340:76–83.
[40] Sievens-Figueroa L, Bhakay A, Jerez-Rozo JI, Pandya N, Romañach RJ, Michniak-Kohn B, Iqbal Z, Bilgili E, Davé RN. Preparation and characterization of hydroxypropyl methyl cellulose films containing stable BCS Class II drug nanoparticles for pharmaceutical applications. Int J Pharm. 2012;423:496–508.
[41] Doub WH, Adams WP, Spencer JA, Buhse LF, Nelson MP, Treado PJ. Raman chemical imaging for ingredient-specific particle size characterization of aqueous suspension nasal spray formulations: a progress report. Pharmaceutical Research. 2007;24:934–45.
[42] Vajna B, Pataki H, Nagy Z, Farkas I, Marosi G. Characterization of melt extruded and conventional Isoptin formulations using Raman chemical imaging and chemometrics. Int J Pharm. 2011;419:107–13.
[43] Rebiere H, Martin M, Ghyselinck C, Bonnet P-A, Brenier C. Raman chemical imaging for spectroscopic screening and direct quantification of falsified drugs. J Pharm Biomed Anal. 2018;148:316–23.
[44] Lanzarotta A, Lorenz L, Batson JS, Flurer C. Development and implementation of a pass/fail field-friendly method for detecting sildenafil in suspect pharmaceutical tablets using a handheld Raman spectrometer and silver colloids. J Pharm Biomed Anal. 2017;146:420–5.

[45] Dégardin K, Desponds A, Roggo Y. Protein-based medicines analysis by Raman spectroscopy for the detection of counterfeits. Forensic Sci Int. 2017;278:313–25.

[46] Breitkreitz MC, Poppi RJ. Trends in Raman chemical imaging. Biomed Spectrosc Imaging. 2012;1:159–83.

[47] Breitenbach J, Schrof W, Neumann J. Confocal Raman-spectroscopy: analytical approach to solid dispersions and mapping of drugs. Pharm Res. 1999;16:1109–3.

[48] Widjaja E, Lee WL, Loo SCJ. Application of Raman microscopy to biodegradable double-walled microspheres. Anal Chem. 2010;82:1277–82.

[49] Kelley WP, Chen S, Floyd PD, Hu P, Kapsi SG, Kord AS, Sun M, Vogt FG. Analytical characterization of an orally-delivered peptide pharmaceutical product. Anal Chem. 2012;84:4357–72.

[50] Slobodan Š. An in-depth analysis of Raman and near-infrared chemical images of common pharmaceutical tablets. Appl Spectrosc. 2007;61:239–50.

[51] Beck R.C.R., et al. 3D printed tablets loaded with polymeric nanocapsules: an innovative approach to produce customized drug delivery systems. Int J Pharm. 2017;528:268–79.

[52] Garvie-Cook H, Frederiksen K, Petersson K, Guy RH, Gordeev S. Characterization of topical film-forming systems using atomic force microscopy and Raman microspectroscopy. Mol Pharm. 2015;12:751–7.

[53] Biggs KB, Balss KM, Maryanoff CA. Pore networks and polymer rearrangement on a drug-eluting stent as revealed by correlated confocal Raman and atomic force microscopy. Langmuir. 2012;28:8238–43.

[54] Breitkreitz MC, Sabin GP, Polla G, Poppi RJ. Characterization of semi-solid self-emulsifying drug delivery systems (SEDDS) of atorvastatin calcium by Raman image spectroscopy and chemometrics. J Pharm Biomed Anal. 2013;73:3–12.

[55] Puncochova K., et al. Non-invasive insight into the release mechanisms of a poorly soluble drug from amorphous solid dispersions by confocal Raman microscopy. Eur J Pharm Biopharm. 2016;101:119–25.

[56] Lee WL, Loei C, Widjaja E, Loo SCJ. Altering the drug release profiles of double-layered ternary-phase microparticles. J of Controlled Release. 2011;151:229–38.

[57] Vukosavljevic, De Kinder L, Siepmann J, Muschert S, Windbergs M. Novel insights into controlled drug release from coated pellets by confocal Raman microscopy. J Raman Spectrosc. 2016;47:757–62.

[58] Windbergs M, Jurna M, Offerhaus HL, Herek JL, Kleinebudde P, Strachan CJ. Chemical imaging of oral solid dosage forms and changes upon dissolution using coherent anti-Stokes Raman scattering microscopy. Anal Chem. 2009;81:2085–91.

[59] Choi DH, Kim KH, Park JS, Jeong SH, Park K. Evaluation of drug delivery profiles in geometric three-layered tablets with various mechanical properties, in vitro–in vivo drug release, and Raman imaging. J of Controlled Release. 2013;172:763–72.

[60] Dong J, Foley JD, Frethem CD, Hoerr RA, Matuszewski MJ, Puskas JE, Haugstad G. Multimodal dynamic imaging of therapeutic biomedical coatings in aqueous medium. Langmuir. 2009;25:5442–5.

[61] Puckett CA, Barton JK. Fluorescein redirects a ruthenium–octaarginine conjugate to the nucleus. J Am Chem Soc. 2009;131:8738–9.

[62] Matthäus C, Kale A, Chernenko T, Torchilin V, Diem M. New ways of imaging uptake and intracellular fate of liposomal drug carrier systems inside individual cells, based on Raman microscopy. Mol Pharm. 2008;5:287–93.

[63] Chernenko T, Sawant RR, Miljkovic M, Quintero L, Diem M, Torchilin V. Raman microscopy for noninvasive imaging of pharmaceutical nanocarriers: intracellular distribution of cationic liposomes of different composition. Mol Pharm. 2012;9:930–6.

[64] Ahlinder L, Ekstrand-Hammarström B, Geladi P, Österlund L. Large uptake of titania and iron oxide nanoparticles in the nucleus of lung epithelial cells as measured by Raman imaging and multivariate classification. Biophys J. 2013;105:310–9.
[65] Kann B, Teubl BJ, Roblegg E, Windbergs M. Label-free in vitro visualization of particle uptake into human oral buccal epithelial cells by confocal Raman microscopy. Analyst. 2014;139:5069–74.
[66] Harvanova MP, Jiravova J, Malohlava J, Tomankova KB, Jirova D, Kolarova H. Raman imaging of cellular uptake and studies of silver nanoparticles effect in BJ human fibroblasts cell lines. Int J Pharm. 2017;528:280–6.
[67] Garrett NL, Lalatsa A, Begley D, Mihoreanu L, Uchegbu L, Schätzlein AG, Moger J. Label-free imaging of polymeric nanomedicines using coherent anti-stokes Raman scattering microscopy. J Raman Spectrosc. 2012;43:681–8.
[68] Ye J, Fox SA, Cudic M, Rezler EM, Lauer JL, Fields GB, Terentis AC. Determination of penetratin secondary structure in live cells with Raman microscopy. J Am Chem Soc. 2010;132:980–8.
[69] Franzen L, Windbergs M. Accessing Raman spectral variability in human stratum corneum for quantitative in vitro depth profiling. J Raman Spectrosc. 2014;45:82–8.
[70] Franzen L, Anderski J, Windbergs M. Quantitative detection of caffeine in human skin by confocal Raman spectroscopy – a systematic in vitro validation study. Eur J Pharmaceutics Biopharmaceutics. 2015;95:110–6.
[71] Mujica Ascencio S, Choe C, Meinke MC, Müller RH, Maksimov GV, Wigger-Alberti W, Lademann J, Darvin ME. Confocal Raman microscopy and multivariate statistical analysis for determination of different penetration abilities of caffeine and propylene glycol applied simultaneously in a mixture on porcine skin ex vivo. Eur J Pharmaceutics Biopharmaceutics. 2016;104:51–8.
[72] Zhang G, Moore DJ, Sloan KB, Flach CR, Mendelsohn R. Imaging the prodrug-to-drug transformation of a 5-fluorouracil derivative in skin by confocal Raman microscopy. J Investigative Dermatol. 2007;127:1205–9.
[73] Pyatski Y, Zhang Q, Mendelsohn R, Flach CR. Effects of permeation enhancers on flufenamic acid delivery in Ex vivo human skin by confocal Raman microscopy. Int J Pharm. 2016;505:319–28.
[74] Mélot M, Pudney PDA, Williamson A-M, Caspers PJ, Van Der Pol A, Puppels GJ. Studying the effectiveness of penetration enhancers to deliver retinol through the stratum cornum by in vivo confocal Raman spectroscopy. J of Controlled Release. 2009;138:32–9.
[75] De Beer T, Burggraeve A, Fonteyne M, Saerens L, Remon JP, Vervaet C. Near infrared and Raman spectroscopy for the in-process monitoring of pharmaceutical production processes. Int J Pharm. 2011;417:32–47.
[76] Simone E, Saleemi AN, Nagy ZK. Application of quantitative Raman spectroscopy for the monitoring of polymorphic transformation in crystallization processes using a good calibration practice procedure. Chem Eng Res Des. 2014;92:594–611.
[77] Pataki H, Csontos I, Nagy ZK, Vajna B, Molnar M, Katona L, Marosi G. Implementation of Raman signal feedback to perform controlled crystallization of Carvedilol. Org Process Res Dev. 2013;17:493–9.
[78] De Beer TR, Baeyens WR, Ouyang J, Vervaet C, Remon JP. Raman spectroscopy as a process analytical technology tool for the understanding and the quantitative in-line monitoring of the homogenization process of a pharmaceutical suspension. Analyst. 2006;131:1137–44.
[79] De Beer TR, Bodson C, Dejaegher B, Walczak B, Vercruysse P, Burggraeve A, Lemos A, Delattre L, Heyden YV, Remon JP, Vervaet C, Baeyens WR. Raman spectroscopy as a process

analytical technology (PAT) tool for the in-line monitoring and understanding of a powder blending process. J Pharm Biomed Anal. 2008;48:772–9.
[80] Nagy B, Farkas A, Gyürkés M, Komaromy-Hiller S, Démuth B, Szabó B, Nusser D, Borbás E, Marosi G, Nagy ZK. In-line Raman spectroscopic monitoring and feedback control of a continuous twin-screw pharmaceutical powder blending and tableting process. Int J Pharm. 2017;530:21–9.
[81] Riolo D, Piazza A, Cottini C, Serafini M, Lutero E, Cuoghi E, Gasparini L, Botturi D, Marino IG, Aliatis I, Bersani D, Lottici PP. Raman spectroscopy as a PAT for pharmaceutical blending: advantages and disadvantages. J Pharm Biomed Anal. 2018;149:329–34.
[82] Walker GM, Bell SEJ, Greene K, Jones DS, Andrews GP. Characterisation of fluidised bed granulation processes using in-situ Raman spectroscopy. Chem Eng Sci. 2009;64:91–8.
[83] Wikström H, Marsac PJ, Taylor LS. In-line monitoring of hydrate formation during wet granulation using Raman spectroscopy. J Pharm Sci. 2005;94:209–19.
[84] Fonteyne M, Gildemyn D, Peeters E, Mortier ST, Vercruysse J, Gernaey KV, Vervaet C, Remon JP, Nopens I, De Beer T. Moisture and drug solid-state monitoring during a continuous drying process using empirical and mass balance models. Eur J Pharm Biopharm. 2014;87:616–28.
[85] Kogermann K, Aaltonen J, Strachan CJ, Pöllänen K, Heinämäki J, Yliruusi J, Rantanen J. Establishing quantitative in-line analysis of multiple solid-state transformations during dehydration. J Pharm Sci. 2008;97:4983–9.
[86] De Beer TR, Vercruysse P, Burggraeve A, Quinten T, Ouyang J, Zhang X, Vervaet C, Remon JP, Baeyens WR. In-line and real-time process monitoring of a freeze drying process using Raman and NIR spectroscopy as complementary process analytical technology (PAT) tools. J Pharm Sci. 2009;98:3430–46.
[87] Knop K, Kleinebudde P. PAT-tools for process control in pharmaceutical film coating applications. Int J Pharm. 2013;457:527–36.
[88] Muller J, Knop K, Thies J, Uerpmann C, Kleinebudde P. Feasibility of Raman spectroscopy as PAT tool in active coating. Drug Dev Ind Pharm. 2010;36:234–43.
[89] Saerens L, Ghanam D, Raemdonck C, Francois K, Manz J, Kruger R, Kruger S, Vervaet C, Remon JP, De Beer T. In-line solid state prediction during pharmaceutical hot-melt extrusion in a 12 mm twin screw extruder using Raman spectroscopy. Eur J Pharm Biopharm. 2014;87:606–15.
[90] Saerens L, Dierickx L, Lenain B, Vervaet C, Remon JP, De Beer T. Raman spectroscopy for the in-line polymer-drug quantification and solid state characterization during a pharmaceutical hot-melt extrusion process. Eur J Pharm Biopharm. 2011;77:158–63.
[91] Li Y-T, Qu L-L, Li D-W, Song Q-X, Fathi F, Long Y-T. Rapid and sensitive in-situ detection of polar antibiotics in water using a disposable Ag–graphene sensor based on electrophoretic preconcentration and surface-enhanced Raman spectroscopy. Biosens Bioelectron. 2013;43:94–100.
[92] Calvet A, Ryder AG. Monitoring cell culture media degradation using surface enhanced Raman scattering (SERS) spectroscopy. Anal Chim Acta. 2014;840:58–67.
[93] Abu-Absi NR, Kenty BM, Cuellar ME, Borys MC, Sakhamuri S, Stachan DJ, Hausladen MC, Li ZJ. Real time monitoring of multiple parameters in mammalian cell culture bioreactors using an in-line Raman spectroscopy probe. Biotechnol Bioeng. 2011;108:1215–21.
[94] Li B, Ray BH, Leister KJ, Ryder AG. Performance monitoring of a mammalian cell based bioprocess using Raman spectroscopy. Anal Chim Acta. 2013;796:84–91.
[95] Andre S, Cristau LS, Gaillard S, Devos O, Calvosa E, Duponchel L. In-line and real-time prediction of recombinant antibody titer by in situ Raman spectroscopy. Anal Chim Acta. 2015;892:148–52.

[96] Pieters S, Vander Heyden Y, Roger JM, D'Hondt M, Hansen L, Palagos B, De Spiegeleer B, Remon JP, Vervaet C, De Beer T. Raman spectroscopy and multivariate analysis for the rapid discrimination between native-like and non-native states in freeze-dried protein formulations. Eur J Pharm Biopharm. 2013;85:263–71.

[97] Angeloni L, Reggente M, Passeri D, Natali M, Rossi M. Identification of nanoparticles and nanosystems in biological matrices with scanning probe microscopy. Wiley Interdiscip Rev Nanomed Nanobiotechnol. 2018;e1521(1939-0041 (Electronic)):–

Markus Lankers

6 Applications in: Environmental Analytics (fine particles)

Abstract: Micro Raman spectroscopy has been applied very early in environmental analytics. However, until now the field of application is quite limited. The main reasons for the low acceptance are high cost of the method and the low throughput. New developments in technology lead to cheaper instrumentation. Automation of Raman microscopy of particles might be a solution for a higher throughput and a broader application in environmental analytics. A more detailed analysis of aerosols and microplastic is good examples that could benefit from this development.

Keywords: micro particles, aerosol, automation, micro plastic, image analysis

6.1 Introduction

The use of a microscope is familiar to most scientific researchers. Raman microscopy (RM) is often seen as an extension of the microscope with the possibility to make more profound statements on the molecular composition of the sample. In environmental analysis, the applications usually go beyond looking at individual small samples and require further quantification. This is also possible with RM, but needs appropriate sample preparation and automation.

In comparison to IR microscopy, RM could be easily incorporated in a conventional optical microscope. The advantages of RM are its high spatial resolution of 1 µm and its insensitivity to water. Since water shows only small Raman bands, water-containing samples can also be analyzed. RM enables the chemical identification of particles and fibers with a spatial resolution of up to 1 µm over a large size range from 1 µm to 5 mm.

One of the biggest disadvantages of RM is the interference with fluorescence from (micro)biological, organic and inorganic impurities, which can make it difficult to identify the particulate matter. The irradiated laser light can cause not only Raman scattering but also fluorescence. Since the latter has significantly higher efficiencies in comparison, the Raman signals can be superimposed by broad fluorescence background. Since most environmental samples possess fluorophores, the avoidance or reduction of fluorescence is of great for all samples used in environmental analytics.

This article has previously been published in the journal *Physical Sciences Reviews*. Please cite as: Lankers, M. Raman Microspectroscopy for Cultural Heritage Studies. *Physical Sciences Reviews* [Online] **2018**, 3. DOI: 10.1515/psr-2018-0178.

https://doi.org/10.1515/9783110515312-006

The measurement parameters can also be optimized to avoid fluorescence (e.g. photobleaching, laser wavelength, laser power, measurement time, objective magnification and confocal mode. In addition, Raman spectroscopy is a comparatively expensive analytical technique compared to IR technology, despite the stormy developments in laser technology, notch filters and detection technology in recent decades.

This might explain the small number of fields of application in the field of environmental analysis. Essentially, two major applications can be identified in environmental analysis:

the study of aerosols and in particular more recently the investigation of microplastic.

Both applications are described in detail below. Particular attention will be paid to the potential for automation of these measurements.

6.2 Applications

6.2.1 Aerosol

Due to the experimental advantages already described, RM was developed for the qualitative analysis of atmospheric aerosols [1, 2]. Here particles with black carbon are of particular importance [3, 4] and/or organic carbon and inorganic materials [5, 6]. Furthermore, Raman spectra were also used for the characterization of atmospheric fungal spores [7], pollen [8, 9] and bacteria [10–12]. RS was also able to study the composition of desert dust [13], marine spray [14] and oceanic aerosols [6]. In particular, the understanding of the chemical reactions in individual aerosol droplets could be promoted with methods of elastic and inelastic light scattering. Raman spectra obtained in these studies were used to classify particles into categories such as bacteria, black carbon, etc. (e.g. [15]). The majority of aerosol studies have been performed on particle collectives, but there are numerous single-particle studies to investigate the properties of carbonaceous aerosols in Europe [16] and of volcanic ash particles [17] and used to identify chemical agent simulants [18]. Numerous methods for automatic measurement and classification of Raman spectra to certain materials have been developed [10].

RS is approximately 10^{12} times weaker than elastic scattering from micrometer-sized particles of the same volume and decreases with the fourth power of the wavelength. The weak intensity of single-particle Raman scattering means that Raman spectral measurements of rapidly moving µm-sized particles are currently not feasible. Extensive measurements of atmospheric aerosol using single-particle RS are not done routinely, because of the time and effort required to collect or trap particles, transfer them if necessary to an RS system, and then measure their Raman spectra (e.g. [5, 16]).

The weakness of Raman scattering means that some method is required to collect (or trap) particles from air, either onto a surface or into a trap that holds the particle in air [19, 20], and then, if necessary, move it to the focal region of the excitation-collection optics [16, 21].

A unique strength of RS is the ability to get structural information out of carbon modifications and is established as a standard characterization technique for any carbon system. Raman spectra of amorphous, nanostructured, diamond–like carbon and nanodiamond could be used to determine structure and composition of carbon films. The carbon microstructure is highly Raman active, making Raman uniquely suited for analysis of carbon materials in different crystal structures. Graphite contains hexagonal planes of carbon atoms, with four carbon atoms in one unit cell [22]. One of the vibrational modes E2g is associated with a Raman peak at 1582 cm^{-1} (G-band) [23]. As a result, graphite with highly single crystallinity only displays a Raman peak at 1582 cm^{-1}. For carbon black material with amorphous microcrystalline structures, another peak around 1350 cm^{-1} (D-band) will also appear. It has been concluded that the peak at 1350 cm^{-1} is attributed to the structure disorder near the edge of the microcrystalline that destructs the structure of the symmetry [23]. Therefore, the Raman peak intensity ratio of ID/IG can be used to characterize the degree of disorder of the graphite materials. It is also discussed that the ID/IG is inversely proportional to the grain size of the carbon black materials for grain sizes larger than 2 nm [24].

The importance for environmental research is the ability to investigate very small carbon particles transported in the atmosphere. The differentiation of carbon particles by their D/G band as well as the possibility to track mineral dust might give the potential to make fingerprints of different sources of atmospheric pollution.

6.2.2 Microplastic

Microplastic (MP) in the oceans or in food is increasingly becoming the focus of public interest. Microplastic particles are either produced directly for use e.g. in cosmetics (primary MP) or may be produced by the decomposition of larger plastic wastes (secondary MP). Reliable studies on the occurrence, uptake and toxicological effects for all higher organisms are currently lacking. Due to their durability, plastics can accumulate in sea water [25–27] as well as in rivers and lakes [28, 29]. Small plastic fragments can be produced from the plastic waste by UV radiation, mechanical and (micro)biological stress. Usually, an upper limit of 5 mm particle size is used, but in various studies 10 mm [30], 2 mm [31], and 1 mm [32] were also used. A generally accepted definition is still pending.

At a certain size, MP particles may even penetrate tissues and cell walls and accumulate in blood vessels with so-far unpredicted consequences. An environmental and human health risk assessment is urgently needed.

A basic requirement for such a risk assessment is reliable data of MPs in the environment. However, these data are not available, particularly for MPs ≤ 500 µm. The quantification of MPs in the smaller size range is connected to many analytical challenges.

The analytical approach to count and identify MP is still in development. Larger MP particles (>500 µm) are usually sorted out manually, while smaller particles are collected on a filter and examined. To ensure identification, the organic matrix must be removed. For purification, the samples can be oxidatively pretreated with Hydrogen Peroxide (H2O2) solutions [29, 33–35] The frequently suggested application of ultrasound should be avoided as smaller fragments are easily produced. This can be avoided under laboratory conditions with the help of a method qualification, but is too cumbersome on the multitude of different plastics for microplastic analysis. The enzymatic purification consists of the successive treatment with a combination of different enzymes, each with specific reaction conditions, which allows this step to continue for a few days. Since each enzyme requires precisely defined reaction conditions, enzymatic processing is generally very time-consuming [36, 37].

The frequently used method of visual classification of MP particles is highly error-prone and leads to the necessity to use additional characterization methods [38]. In a study by Nielsen [39, 40], particles were both visually classified and characterized by Raman-spectroscopy. Only 68 % of all particles could be confirmed spectroscopically as microplastics. The analysis of fibers showed a similar trend with a total of 75 % correctly allocated samples, underlining the need for an identification method especially for small particles.

RM and FTIR are the most commonly used identification techniques to MP particles down to the lower-micrometer range are Raman and Fourier transform infrared (FTIR) spectroscopy. These two techniques have different advantages and disadvantages.

RM has a size resolution (detects particles down to a size of 1 µm), which is superior compared to FTIR but fluorescence often compromises the quality of Raman spectra of environmental MPs. In contrast, FTIR has a less precise size resolution (detects particles down to a size of 10 20 µm), and its spectral quality is not influenced by fluorescence but by the presence of water.

RM has been used in a number of studies on the occurrence of MP in water [27, 35, 39, 41]. In most cases, however, only large particles or a small sample of all particles were examined using RM. Large MP particles (>500 µm) can be spectroscopically examined directly under the Raman microscope. Smaller particles must be transferred to a filter or other suitable sample carrier and analyzed. The filtration and also the selection of the particles to be measured and finally the measurement and evaluation is generally done manually [39, 40]. If one estimates 1 min for the selection and focusing, as well as for the measurement and evaluation of the results, one could expect a total measuring time per particle of approx. 4–5 min. In one working day, the analysis of approx. 100 particles is possible which is a small number compared to several thousand particles to be analyzed in one sediment sample. A visual assessment of the particles is not reliable and like all methods of visual classification depends on the operator. However, a corresponding classification can be carried out in the range of a few seconds. The specificity of RM as well as the spatial resolution is superior compared with visual classification but slows down the analysis by an order of magnitude.

An approach to reduce this gap is automation and will be discussed in more detail in the next chapter.

6.3 Automation

The majority of Raman instrument manufacturer produces instruments for the scientific market. These instruments are mainly multipurpose instruments based on a microscope. However, over the last ten years a number of companies introduced instruments specially equipped for particle identification. These instruments are able to detect and classify particles and finally identify these detected particles by means of Raman spectroscopy. Eventually these Raman spectra can be identified automatically by data base search. This approach reduces the time for an analysis of 100 hundred particles from 500 min to 120 min. Considering a 24 h working day for an automated machine using sample changer an identification rate of 1000 particles/ day can be reached. The first instrument which hit the market was the Single Particle Explorer by rap.ID (now Unchainedlabs). This instrument is an automated Raman microscope combined with the unique feature of LIPS microscopy gathering elemental information. Malvern Instruments offers the morphology G4 using the same general idea of particle identification. Finally, Horiba introduced a software which is able to retrofit Particle ID capabilities to existing Horiba Raman microscopes. Nowadays, all major vendors of RM offer particle identification features to some extent.

6.3.1 Substrate and sampling

The first step in automated RM is the proper preparation of the sample. Particles are placed by filtration or impaction on an appropriate sample surface.

When selecting a Raman substrate, it is important to consider both the physical and optical properties of the material, as these influence sample preparation, spectral acquisition and subsequent data processing [9]. Especially for measurements with a large amount of data, such as automated measurements, the subsequent data processing plays a decisive role. The background signal of the substrate depends on the material and should be minimized to the superposition of Raman signals by background signals [42].

Raman substrates are often used to improve signals (surface enhanced Raman scattering SERS) by many orders of magnitude [43, 44]. The most commonly used surfaces for SERS are silver and gold, which exhibit plasmon resonances in the visible to near infrared range, depending on the structure of the nanoscale material [43].

The currently most widely used substrate in the study of cells is CaF2, as the material is largely transparent, has low absorption and shows only a Raman line at 321 cm^{-1} combined with a remarkable low background signal. Unfortunately, the use of CaF2 is limited for cost sensitive applications due to its high price of $10. The material also allows FTIR examinations and Raman on the same sample.

Quartz and standard glasses, which are commonly used in microscopy, have a more intensive and, above all, wider background. In the past, metallic and metallized sample carriers were often used for examinations. Gold and aluminum coated filters or even silver filters have often been proposed in forensic analysis as Raman substrates for particle separation and subsequent investigation. The metal substrate and metallized surfaces have no vibrational background. Electrolytically deposited nickel was used as a substrate for cell investigations. Due to the high purity, no interfering signals were found.

Especially in the field of environmental analysis, the break-proof and inexpensive metal substrates represent an interesting approach [45]. An ideal background should provide a good signal-to-noise ratio and spatial resolution. Metal should give no background signal in Raman spectroscopy. But due to the production process background signals could be quite different. Figure 6.1 shows the background signal for different materials. A number of studies have investigated the Raman properties of metallic and metal-coated Raman substrates in biological analyses [43]. Previous work has shown that polished stainless steel has the spectral intensity and reproducibility as the laser photons are reflected by the polished metal surface [15, 46]

6.3.2 Automation workflow

RM is very well suited for automation, as essential elements from the automation of microscopes can be transferred: Over the last few years, some Raman instrument manufacturers have added special modes to their instruments to automatically measure particles. Even if this software has very different manufacturers, the actual workflow is the same for you. It is shown in Figure 6.2.

After sampling a surface, the surface is captured with a number of microscopic images. The images are segmented in a second step and the resulting size and morphological properties of the particles are recorded. Particles or particle groups can finally be selected for measurement on the basis of the morphological data. The measurement is done by point or mapping measurements at the coordinates of the selected particles. Finally, the spectra are analyzed in an evaluation step. Different strategies can be applied

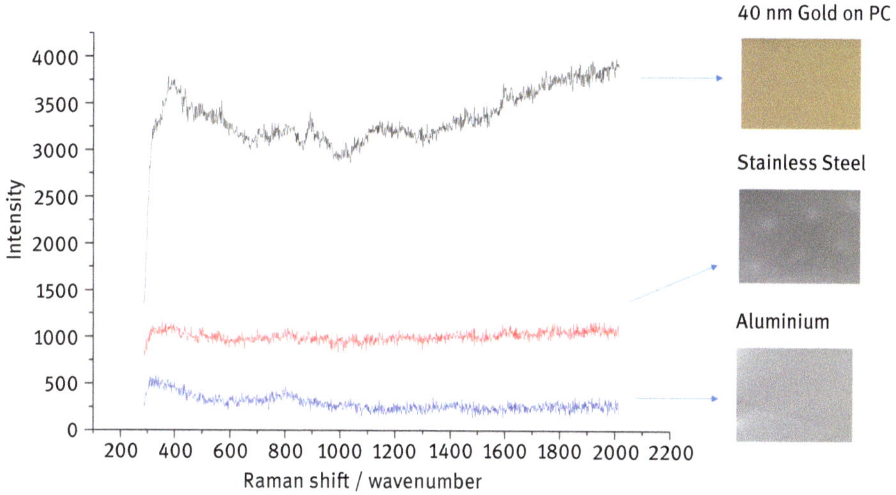

Figure 6.1: Metal should give no background signal in Raman spectroscopy. But due to the production process background signals could be quite different. Signal A is a polycarbonate membrane with a 40 nm gold coating. Using thicker coatings background signal decreases. Signal B is derived from a electropolished steel surface. Surface C shows background signal from a 99,99 aluminum surface. The background is very low.

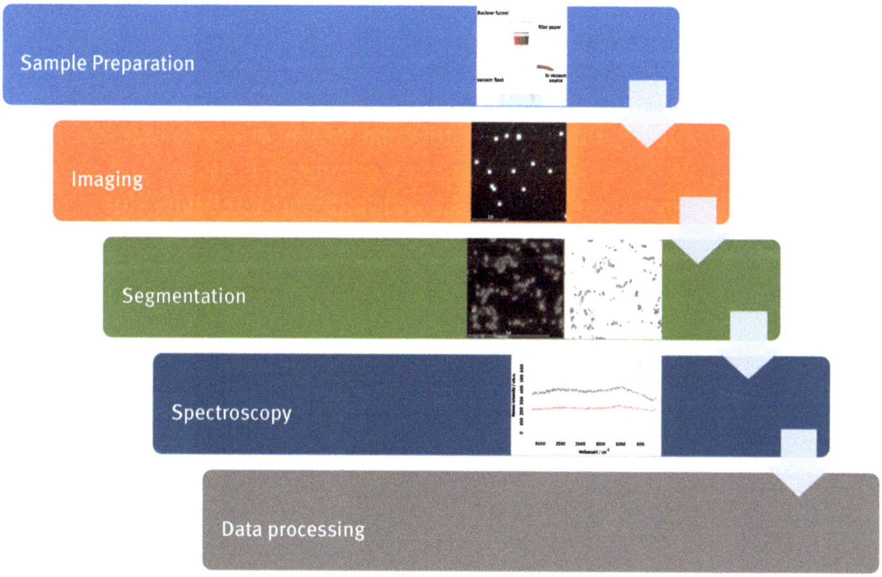

Figure 6.2: The general workflow for automated particle identification: sampling by filtration of powder deposition, microscopic imaging of the sample surface, particle recognition by segmentation, Raman spectroscopy of the single particles and finally data processing and evaluation of the spectroscopic result.

- Multivariate analysis
- Spectrum analysis with database search
- Classifications by spectral characteristics

All methods have their advantages and disadvantages and must be selected on a case-by-case basis.

Even though this workflow is common to all methods, there are considerable differences in system integration. The selection of different lighting methods is fundamental. In particular, the use of darkfield illumination extends the application in the area of the analysis of transparent and very small particles. The use of different optics for particle detection and spectroscopy can lead to significantly shorter measurement times or to a significantly improved signal-to-noise ratio. A punctual autofocus system pays off especially with large particles.

6.3.3 Parameterization / validation

In the classical application of RM, the user performs a single analysis and, depending on the result, changes the measurement time or the laser power of the measurement in order to achieve a better result in a repeat measurement. This is only possible to a limited extent during a long measurement of thousands of particles. For this reason, method development and/or validation is very important for automated measurements. In addition to the illumination, which influences the segmentation of the particles and thus their detection performance, a careful selection of the laser power and exposure time is necessary. A number of different measurement steps are performed in automated Raman spectroscopic identification. An example of the influence of a single parameter is shown in Figure 6.3. 2 µm polystyrene spheres are illuminated with different light intensities. Due to the stronger light scattering particles appear larger in the images. The recognized sizes by the image analysis increases from 2 to 8 µm and might lead to a wrong particle selection for the spectroscopic investigation.

6.3.4 Advantages over other techniques / future trends

There are only a few publications describing the applications of automatic Raman spectroscopy in environmental science. The main applications of this technique are mainly in the field of industrial forensics for contamination control and the investigating the ratio of different formulation components in the field of pharmaceutical quality control [47–50]. The identification of bacteria with automated Raman systems is also described.

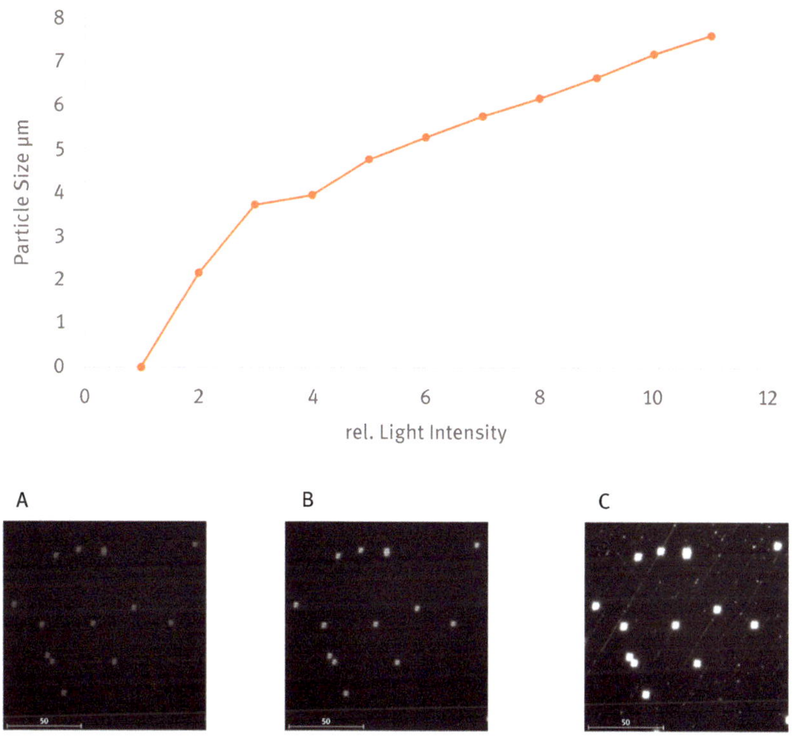

Figure 6.3: A number of different measurements performed in automated Raman spectroscopy. Not only the wavelength calibration is important but also the parametrization of segmentation, illumination and Light intensity. 2-µm polystyrene spheres are illuminated with different light intensity in darkfield illumination. The size recognized by the image analysis increases from 2 to 8 µm.

Recently, a couple of publications [49, 50] have been published discussing the use of automated RM for the identification of MP. Gerdts et al. [51] compare the use of automated Raman spectroscopy with FTIR imaging for the analysis of microplastics. While micro Raman spectroscopy is expected to be characterized by a higher dynamic range of particle sizes (identification from 1 µm possible, as well as a better yield of identifications in the range of 10–20 µm), FTIR imaging can be advantageous due to its ability to characterize plastic particle agglomerates. Due to the strong tendency of microplastic particles to fluoresce, Raman analysis can only be performed with a relatively long measurement time.

Schymanski et al. [52] investigated the distribution of microplastics with the help of automatic RA in different mineral and drinking water samples filled into different containers. The automated technique made it possible to quantify a large number of samples quantitatively with regard to microparticles in the *x*–*y* µm range. Interestingly, glass bottles were found to be more contaminated with microparticles than plastic bottles or tetrapacks. The confusing interference of Raman spectroscopic analysis of plastic samples by pigments was discussed as a weak point of the analysis. The Raman spectra of dyes are often strongly amplified by the resonance Raman effect and often mask the spectrum of the actual plastic. This is also a problem that frequently occurs in the forensics of fibers. Since this effect naturally depends on the excitation wavelength, the use of dye databases for the corresponding wavelength can help and the use of long-wave excitation lasers minimize the problem. At this point automated RM might be the only technique investigated which is able to count and characterize MP in the small size range 1–10 µm reliably. However, more research is needed to overcome the disturbing interference of fluorescence.

An alternative approach of particle identification can be found in Raman Imaging which is also widely used in cell characterization. Alternatively, increasingly stimulated Raman spectroscopy [53]and CARS [54] are used as labelfree imaging methods. Due to the large sample sizes that arise in the described applications, these methods are currently still too time and cost intensive to be used for similar investigations.

Hill et al. [55] published a study of airborne particles using the REBS "Resource Effective Bioidentification System" (REBS) developed by the Batelle Institute. This REBS-based RM can be used as a line scan Raman imaging spectrometer device detecting particles on a metallized tape. The particles are deposited on the tape in the same device *vy* means of a vacuum pump. The spectra show for different materials, e.g. polystyrene, black carbon and several other materials easily distinguishable fingerprints.

Raman spectra from a predetermined time interval e.g.15-min environmental sample (approximately 35–50 particles, 158 spectra) were analyzed using a hierarchical clustering technique. The cluster spectra were found to be consistent with soot, inorganic aerosol, and other organic compounds.

During a 7-h period sample campaign, it was possible to measure 32,718 spectra. 5892 could be analyzed by clustering. A total of 95 % of these spectra could be attributed to soot. The number of soot particles exhibiting the amorphous carbon D-G bands recorded over time increases during morning and then decreases. The other detected categories do not exhibited this temporal change. These data gives an idea about the potential of the automated RM to measure thousands of time-resolved aerosol Raman spectra. More extensive RS-based studies of atmospheric aerosols should be expected in the future. However, there is still some basic research needed to evaluate the interaction of organic material with slightly absorbing organic material which can be easily charred by the laser beam, Furthermore the cross-section of carbon particles for visible light is huge. A small carbon particle sitting on a larger particle could mask the particle as soot leading to higher soot populations. The use of a longer wavelength could correct the problem.

References

[1] Rosen H, Novakov T. Raman scattering and the characterisation of atmospheric aerosol particles. Nature. 1977;266:708–10.
[2] Sinanis S, Aleksandrova M, Schaber K. Characterization of multicomponent aerosols by Raman spectroscopy. Aerosol Sci Technol. 2011;45:741–7.
[3] Sadezky A, Muckenhuber H, Grothe H, Niessner R, Pöschl U. Raman microspectroscopy of soot and related carbonaceous materials: spectral analysis and structural information. Carbon N Y. 2005;43:1731–42.
[4] Grafen M, Schweiger G, Esen C, Ostendorf A. Time-resolved measurement of elemental carbon in urban environment: comparison of Raman backscattering and aethalometer results. J Aerosol Sci. 2018;117:34–43.
[5] Deng C, Brooks SD, Vidaurre G, Thornton DCO. Using Raman microspectroscopy to determine chemical composition and mixing state of airborne marine aerosols over the pacific ocean. Aerosol Sci Technol. 2014;48:193–206.
[6] Gaffney JS, Marley NA, Smith KJ. Characterization of fine mode atmospheric aerosols by Raman microscopy and diffuse reflectance FTIR. J Phys Chem A. 2015;119:4524–32.
[7] Ghosal S, Macher JM, Ahmed K. Raman microspectroscopy-based identification of individual fungal spores as potential indicators of indoor contamination and moisture-related building damage. Environ Sci Technol. 2012;46:6088–95.
[8] Guedes A, Ribeiro H, Fernández-González M, Aira MJ, Abreu I. Pollen Raman spectra database: application to the identification of airborne pollen. Talanta. 2014;119:473–8.
[9] Schulte F, Lingott J, Panne U, Kneipp J. Chemical characterization and classification of pollen. Anal Chem. 2008;80:9551–6.
[10] Ronningen TJ, Schuetter JM, Wightman JL, Murdock A, Bartko AP. Raman spectroscopy for biological identification [Internet]. In: Paul Schaudies (ed.). Biological Identification. Elsevier, 2014: 313–33.
[11] Vanoni M, Vai M, Popolo L, Alberghina L. Structural heterogeneity in populations of the budding yeast saccharomyces cerevisiae. J Bacteriol. 1983;156:1282–91.
[12] Tripathi A, Jabbour RE, Guicheteau JA, Christesen SD, Emge DK, Fountain AW, et al. Bioaerosol analysis with Raman chemical imaging microspectroscopy. Anal Chem. 2009;81:6981–90.
[13] Sobanska S, Hwang H, Choël M, Jung H, Eom H, Kim H, et al. Investigation of the chemical mixing state of individual Asian dust particles by the combined use of electron probe X-ray microanalysis and Raman microspectrometry. Anal Chem. 2012;84:3145–54.
[14] Gibson ER, Hudson PK, Grassian VH. Physicochemical properties of nitrate aerosols: implications for the atmosphere. J Phys Chem A. 2006;110:11785–99.
[15] Hiranuma N, Brooks SD, Gramann J, Auvermann BW. High concentrations of coarse particles emitted from a cattle feeding operation. Atmos Chem Phys. 2011;11:8809–23.
[16] Ivleva NP, McKeon U, Niessner R, Pöschl U. Raman microspectroscopic analysis of size-resolved atmospheric aerosol particle samples collected with an ELPI: soot, humic-like substances, and inorganic compounds. Aerosol Sci Technol. 2007;41:655–71.
[17] Ivleva NP, Huckele S, Weinzierl B, Niessner R, Haisch C, Baumann T, et al. Identification and characterization of individual airborne volcanic ash particles by Raman microspectroscopy. Anal Bioanal Chem. 2013;405:9071–84.
[18] Aggarwal RL, Di Cecca S, Farrar LW, Shabshelowitz A, Jeys TH. Sensitive detection and identification of isovanillin aerosol particles at the pg/cm^3 mass concentration level using Raman spectroscopy. Aerosol Sci Technol. 2015;49:753–6.

[19] Vehring R, Aardahl CL, Schweiger G, Davis EJ. The characterization of fine particles originating from an uncharged aerosol: size dependence and detection limits for Raman analysis. J Aerosol Sci. 1998;29:1045–61.
[20] Pan Y-L, Hill SC, Coleman M. Photophoretic trapping of absorbing particles in air and measurement of their single-particle Raman spectra. Opt Express. 2012;20:5325.
[21] Kalasinsky KS, Hadfield T, Shea AA, Kalasinsky VF, Nelson MP, Neiss J, et al. Raman chemical imaging spectroscopy reagentless detection and identification of pathogens: signature development and evaluation. Anal Chem. 2007;79:2658–73.
[22] Reich S, Thomsen C. Raman spectroscopy of graphite. Philos Trans R Soc A Math Phys Eng Sci. 2004;362:2271–88.
[23] Yan Wang DC. Alsmeyer and RLM. Raman spectroscopy of carbon materials: structural basis of observed spectra. Chem Mater. 1990;2:557–63.
[24] Tuinstra F, Koenig JL. Raman spectrum of graphite. J Chem Phys. 1970;53:1126–30.
[25] Ivar Do Sul JA, Costa MF. The present and future of microplastic pollution in the marine environment. Environ Pollut. 2014;185:352–64.
[26] Gregory MR. Environmental implications of plastic debris in marine settings–entanglement, ingestion, smothering, hangers-on, hitch-hiking and alien invasions. Philos Trans R Soc B Biol Sci. 2009;364:2013–25.
[27] Cole M, Lindeque P, Halsband C, Galloway TS. Microplastics as contaminants in the marine environment: A review. Mar Pollut Bull. 2011;62:2588–97.
[28] Eerkes-Medrano D, Thompson RC, Aldridge DC. Microplastics in freshwater systems: A review of the emerging threats, identification of knowledge gaps and prioritisation of research needs. Water Res. 2015;75:63–82.
[29] Dris R, Imhof H, Sanchez W, Gasperi J, Galgani F, Tassinand B, et al. Beyond the ocean: contamination of freshwater ecosystems with (micro-) plastic particles. Environ Chem. 2015;12:539–50.
[30] Graham ER, Thompson JT. Deposit- and suspension-feeding sea cucumbers (Echinodermata) ingest plastic fragments. J Exp Mar Bio Ecol. 2009;368:22–9.
[31] Ryan PG, Moore CJ, Van Franeker JA, Moloney CL. Monitoring the abundance of plastic debris in the marine environment. Philos Trans R Soc B Biol Sci. 2009;364:1999–2012.
[32] Browne MA, Crump P, Niven SJ, Teuten E, Tonkin A, Galloway T, et al. Accumulation of microplastic on shorelines woldwide: sources and sinks. Environ Sci Technol. 2011;45:9175–9.
[33] Nuelle M-T, Dekiff JH, Remy D, Fries E. A new analytical approach for monitoring microplastics in marine sediments. Environ Pollut. 2014;184:161–9.
[34] Tagg AS, Sapp M, Harrison JP, Ojeda JJ. Identification and quantification of microplastics in wastewater using focal plane array-based reflectance Micro-FT-IR imaging. Anal Chem. 2015;87:6032–40.
[35] Imhof HK, Laforsch C, Wiesheu AC, Schmid J, Anger PM, Niessner R, et al. Pigments and plastic in limnetic ecosystems: A qualitative and quantitative study on microparticles of different size classes. Water Res. 2016;98:64–74.
[36] Löder MGJ, Gerdts G. Methodology used for the detection and identification of microplastics —a critical appraisal [Internet]. In: Bergmann M, Gutow L, Klages M. (eds). Marine anthropogenic litter. Cham: Springer International Publishing, 2015:201–27.
[37] Cole M, Webb H, Lindeque PK, Fileman ES, Halsband C, Galloway TS, et al. Isolation of microplastics in biota-rich seawater samples and marine organisms. Sci Rep. 2014;4:4528.
[38] Remy F, Collard F, Gilbert B, Compère P, Eppe G, Lepoint G, et al. When microplastic is not plastic: the ingestion of artificial cellulose fibers by macrofauna living in seagrass macrophytodetritus. Environ Sci Technol. 2015;49:11158–66.

[39] Enders K, Lenz R, Stedmon CA, Nielsen TG. Abundance, size and polymer composition of marine microplastics ≥10μm in the Atlantic Ocean and their modelled vertical distribution. Mar Pollut Bull. 2015;100:70–81.
[40] Lenz R, Enders K, Stedmon CA, MacKenzie DMA, Nielsen TG. A critical assessment of visual identification of marine microplastic using Raman spectroscopy for analysis improvement. Mar Pollut Bull. 2015;100:82–91.
[41] Van Cauwenberghe L, Vanreusel A, Mees J, Janssen CR. Microplastic pollution in deep-sea sediments. Environ Pollut. 2013;182:495–9.
[42] Mikoliunaite L, Rodriguez RD, Sheremet E, Kolchuzhin V, Mehner J, Ramanavicius A. et al. The substrate matters in the Raman spectroscopy analysis of cells. Sci Rep. 2015;5:1–10.
[43] Domke KF. Surface enhanced Raman spectroscopy. analytical, biophysical and life science applications, 2011. Edited by Sebastian Schlücker. [Internet]. Weinheim: WILEY-VCH.
[44] Ramoji A, Galler K, Glaser U, Henkel T, Mayer G, Dellith J, et al. Characterization of different substrates for Raman spectroscopic imaging of eukaryotic cells. J Raman Spectrosc. 2016;47:773–86.
[45] Lewis AT, Gaifulina R, Isabelle M, Dorney J, Woods ML, Lloyd GR, et al. Mirrored stainless steel substrate provides improved signal for Raman spectroscopy of tissue and cells. J Raman Spectrosc. 2017;48:119–25.
[46] Oßmann BE, Sarau G, Schmitt SW, Holtmannspötter H, Christiansen SH, Dicke W, et al. Development of an optimal filter substrate for the identification of small microplastic particles in food by micro-Raman spectroscopy. Anal Bioanal Chem. 2017;409:4099.
[47] Kinnunen H, Hebbink G, Peters H, Huck D, Makein L, Price R, et al. Extrinsic lactose fines improve dry powder inhaler formulation performance of a cohesive batch of budesonide via agglomerate formation and consequential co-deposition. Int J Pharm. 2015;478:53–9.
[48] Fischer S, Valet O, Lankers M. Raman Microscopy for characterization of particles. In: Mahler H, Jiskoot W (eds.). Analysis of Aggregates and Particles in Protein Pharmaceuticals 2012.
[49] Käppler A, Fischer D, Oberbeckmann S, Schernewski G, Labrenz M, Eichhorn KJ, et al. Analysis of environmental microplastics by vibrational microspectroscopy: FTIR, Raman or both? Anal Bioanal Chem. 2016;408:8377–91.
[50] Elert AM, Becker R, Duemichen E, Eisentraut P, Falkenhagen J, Sturm H, et al. Comparison of different methods for MP detection: what can we learn from them, and why asking the right question before measurements matters? Environ Pollut. 2017;231:1256–64.
[51] Cabernard L, Roscher L, Lorenz C, Gerdts G, Primpke S. Comparison of Raman and Fourier transform infrared spectroscopy for the quantification of microplastics in the aquatic environment. Environ Sci Technol. 2018;52:13279–88.
[52] Schymanski D, Goldbeck C, Humpf HU, Fürst P. Analysis of microplastics in water by micro-Raman spectroscopy: release of plastic particles from different packaging into mineral water. Water Res. 2018;129:154–62.
[53] Zada L, Leslie HA, Vethaak AD, Tinnevelt GH, Jansen JJ, de Boer JF, et al. Fast microplastics identification with stimulated Raman scattering microscopy. J Raman Spectrosc. 2018;49:1136–44.
[54] Cole M, Lindeque P, Fileman E, Halsband C, Goodhead R, Moger J, et al. Microplastic ingestion by zooplankton. Environ Sci Technol. 2013;47:6646–55.
[55] Doughty DC, Hill SC. Automated aerosol Raman spectrometer for semi-continuous sampling of atmospheric aerosol. J Quant Spectrosc Radiat Transf. 2017;188:103–17.

Christoph Krafft and Jürgen Popp

7 Micro-Raman spectroscopy in medicine

Abstract: A potential role of optical technologies in medicine including micro-Raman spectroscopy is diagnosis of bacteria, cells and tissues which is covered in this chapter. The main advantage of Raman-based methods to complement and augment diagnostic tools is that unsurpassed molecular specificity is achieved without labels and in a nondestructive way. Principles and applications of micro-Raman spectroscopy in the context of medicine will be described. First, Raman spectra of biomolecules representing proteins, nucleic acids, lipids and carbohydrates are introduced. Second, microbial applications are summarized with the focus on typing on species and strain level, detection of infections, antibiotic resistance and biofilms. Third, cytological applications are presented to classify single cells and study cell metabolism and drug–cell interaction. Fourth, applications to tissue characterization start with discussion of lateral resolution for Raman imaging followed by Raman-based detection of pathologies and combination with other modalities. Finally, an outlook is given to translate micro-Raman spectroscopy as a clinical tool to solve unmet needs in point-of-care applications and personalized treatment of diseases.

Keywords: Biomedical spectroscopy, biophotonics, microbiology, cytology, histopathology

7.1 Introduction

7.1.1 Motivation for micro-Raman spectroscopy in medicine

In a general sense, medicine is the science and practice of the diagnosis, treatment and prevention of disease. Pathology which is a major field in modern medicine and diagnosis deals with the causal study of disease, in other words "the study of paths" (meaning of the Greek roots of the term pathology) by which disease comes. A potential role of optical technologies in medicine including micro-Raman spectroscopy is diagnosis of bacteria, cells and tissues which is covered in this chapter. The perspectives, potentials and trends of *ex vivo* and *in vivo* optical molecular pathology have recently been described [1]. A number of optical techniques have been suggested in molecular pathology such as fluorescence of autofluorophores or fluorescent markers, narrow band imaging, hyperspectral imaging, photoacoustic tomography

This article has previously been published in the journal *Physical Sciences Reviews*. Please cite as: Krafft, C., Popp, J. Micro-Raman spectroscopy in medicine *Physical Sciences Reviews* [Online] **2019**, 4. DOI: 10.1515/psr-2017-0047.

https://doi.org/10.1515/9783110515312-007

and optical coherence tomography (OCT) that all depend on absorption, emission or scattering of light. The main advantage of Raman-based methods to complement and augment diagnostic tools in pathology is that unsurpassed molecular specificity is achieved without labels and in a nondestructive way. Principles and applications of micro-Raman spectroscopy in the context of medicine will be described here.

7.1.2 Raman spectra of biomolecules

The Raman spectra of bacteria, cells and tissues constitute a superposition of spectral contributions from the underlying biomolecules. The main classes of biomolecules in bacteria, cells and soft tissues are proteins, nucleic acids, lipids and carbohydrates. They are composed by a limited number of defined subunits such as amino acids, nucleobases, fatty acids and saccharides. It goes beyond the scope of this chapter to introduce the Raman spectra of all subunits. They can be found in the literature [2]. Figure 7.1 shows representative Raman spectra of proteins, nucleic acids, lipids and carbohydrates to demonstrate the concepts and the challenges of micro-Raman spectroscopy in medicine. The most prominent challenge is that each subunit and, consequently, also each biomolecule have relatively weak Raman cross-sections, and successful detection in experiments using spontaneous Raman spectroscopy (without signal enhancement) requires sensitive instrumentation and significant concentrations of each component. Exceptions are biomolecules with chromophores whose cross-sections are enhanced at excitation wavelengths near electronic absorption due to a resonance effect. Examples are the prosthetic heme group in hemoglobin and cytochrome proteins at visible excitation, and aromatic amino acids and nucleobases at ultraviolet excitation. Hard tissues contain in addition calcified components such as hydroxyapatite or calcium oxalate which also give relatively intense Raman signals.

Raman spectra of proteins concanavalin A (A) and bovine serum albumin (B) are shown in Figure 7.1. Main Raman bands of proteins are assigned to aromatic amino acids phenylalanine (621, 1,003, 1,031 and 1,208 cm^{-1}), tyrosine (643, 829 and 853 cm^{-1}), tryptophan (759 and 1,555 cm^{-1}), C–C groups (1,126 cm^{-1}), S–S groups of disulfide bonds (508 and 550 cm^{-1}), CH_2/CH_3 groups of aliphatic amino acids (1,340 and 1,449 cm^{-1}), the amide vibrations (1,235–1,320 and 1,630–1,690 cm^{-1}) and C–C_α vibrations (939 cm^{-1}) of the peptide backbone. In general, bands depend on the primary (amino acid sequence), secondary (alpha helix, beta sheets etc.) and tertiary (three-dimensional) structure, the hydrophilic or hydrophobic environment and interaction with other biomolecules. Therefore, the Raman spectrum of each protein constitutes a specific fingerprint. As thousands of different proteins might be present in a biological sample, the Raman spectral contributions of all proteins overlap and the signatures of individual proteins cannot be resolved anymore.

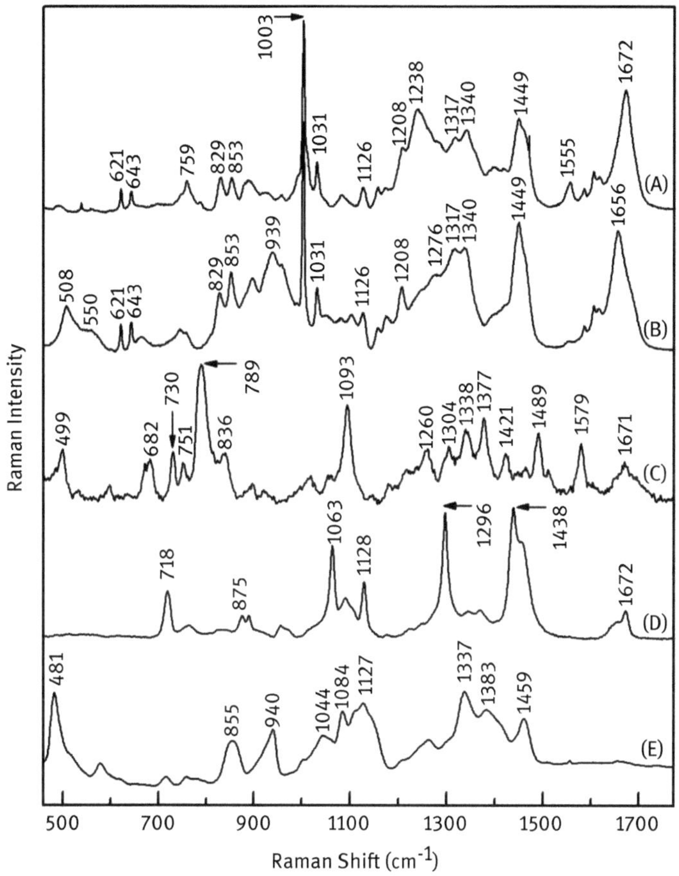

Figure 7.1: Reference Raman spectra of proteins concanavalin A (A), bovine serum albumin (B), deoxyribonucleic acid (C), the lipid sphingomyelin (D) and glycogen (E).

A Raman spectrum of deoxyribonucleic, DNA, (C) is included in Figure 7.1. Main Raman bands of nucleic acids are assigned to the phosphate backbone (810–840 and 1,090–1,100 cm^{-1}) and nucleobases adenine, A, (730, 1,304 and 1,338 cm^{-1}), thymine, T, (499, 751, 1,377 and 1,671 cm^{-1}), guanine, G, (682, 1,489 and 1,579 cm^{-1}), cytosine, C, (789 and 1,260 cm^{-1}). Careful inspection reveals more bands, whose assignments can be found in the literature [2]. Raman bands depend on the nucleotide sequence and backbone geometry that can be A-form in RNA, B-form in DNA and for some oligonucleotides at high salt concentration DNA adopts a Z-form configuration. Upon dehydration, the regular B-DNA conformation in cell nuclei transforms first to an A-like conformation and then to a less ordered, denatured conformation. In the condensed state, DNA is highly packed in the nucleoli of the cell nucleus. During cell division and proliferation, the DNA is in a more uncondensed state and

DNA Raman bands are more intense although the DNA content is identical. As the human genome consists of more than 3 billion bases pairs, changes in few bases in gene sequences that are associated with cancer or other diseases are unlikely to be detected by Raman spectroscopy. Such spectral variations are simply too small. However, the genome determines the phenotype of bacteria, cells and tissues, the proliferation rate, expression of proteins or accumulation of lipids and metabolites, and such phenotypic markers can be detected as Raman spectroscopic signatures for classification and disease recognition.

In biology, lipids are substances of biological origin that are soluble in nonpolar solvents. They comprise a group of naturally occurring molecules that include fats, waxes, sterols, fat-soluble vitamins, mono-, di- and triglycerides, phospholipids and other. Triglycerides are stored in adipose tissue as long-term energy reserve. Main Raman bands of the lipid sphingomyelin (Figure 7.1D) are assigned to the hydrophobic head group choline (716 and 875 cm^{-1}), the phosphate group (1,080 cm^{-1}), the hydrophobic saturated fatty acids chains (1,063, 1,128, 1,296 and 1,438 cm^{-1}) and the ceramide backbone (1,672 cm^{-1}). Other typical lipids bands, that are not evident in Figure 7.1, are due to unsaturated fatty acids chains near 1,270 and 1,660 cm^{-1} and due to ester groups at the linkage between the glycerol backbone and fatty acid chains near 1,740 cm^{-1}.

The Raman spectra of carbohydrates show a large degree of variability because different monosaccharides as subunits are known such as glucose, fructose and galactose. Glycogen is a multi-branched polysaccharide of glucose that serves as a form of energy storage in organisms. The Raman spectrum of glycogen in Figure 7.1E shows a number of broad bands with maxima near 481, 855, 940, 1,044, 1,084, 1,127, 1,337, 1,383 and 1,459 cm^{-1}.

7.2 Micro-Raman spectroscopy of bacteria

7.2.1 Experimental considerations

The fundamentals of vibrational spectroscopies (Raman and infrared spectroscopy) in medical microbiology were already reviewed some time ago [3]. Micro-Raman spectroscopy enables to focus the excitation laser down to about 1 µm in diameter, which is on the order of magnitude of the size of bacteria. Therefore, micro-Raman-spectroscopy can probe single bacteria which makes the time-consuming cultivation unnecessary and reduces the amount of potentially hazardous biomaterial during preparation. Bacteria in complex matrices or body fluids need to be extracted before. Several approaches for Raman-compatible isolation of bacterial cells were summarized [4]. Raman spectra can directly be collected from microbial colonies on culture plates. The preparation procedure that was developed for infrared spectroscopy can also be applied for Raman spectroscopy: bacteria are harvested from colonies on culture plates, transferred to centrifuge tubes, washed with buffer, the supernatant is discarded, the pellet is resuspended with buffer and the pellet is pipetted on a Raman-compatible substrate. Whereas glass substrates are appropriate with green excitation lasers, more expensive substrates made of quartz or calciumfluoride are required for Raman spectroscopy using near-infrared excitation lasers between 700 and 850 nm. Different substrates were characterized for Raman spectroscopy of eukaryotic cells [5] that are also appropriate for prokaryotic cells.

7.2.2 Microbial applications of micro-Raman spectroscopy

While some bacteria can cause serious infectious diseases, others are vital parts of the gut flora or live on the skin; other bacteria can spoil food, while other strains are utilized in food, pharmaceutical and chemical industry to produce cheese, vinegar, yoghurt, antibiotics, hormones, lactic acids and many other products. Fast and reliable identification methods are needed to distinguish the useful and harmless bacteria from the unwanted and toxic ones and to provide an efficient and appropriate treatment of infections. In times of growing antibiotic resistances, it is furthermore desirable to identify new target structures for innovative and effective drugs.

7.2.2.1 Typing on species and strain level

In medicine, the exact identification of bacteria is necessary in order to provide the appropriate therapy and to prevent antibiotic resistance which may further delay administration of the most appropriate narrow-spectrum antibiotic. Traditional microbiological identification and infection diagnoses are time-consuming and labor-intensive processes: at least 12 h up to 24 h incubation is required to obtain an accurate colony count. Slowly growing bacteria require even longer incubation. An additional 24 ± 12 h is needed for organism identification and susceptibility testing. As a first differentiation criteria, relatively unspecific morphological parameters, such as size, shape and color of the bacteria as well as of the colonies, are used which need to be complemented with expensive and tedious metabolic tests, such as the ability to grow in various media under different conditions, degradation of certain substrates and enzyme activity.

Microbial cells from different species or strains vary in their chemical composition, e. g. in the concentration, structure and type of proteins, carbohydrates, lipids and DNA/RNA sequences. These variations can be monitored by Raman spectroscopy and Raman spectra provide complex and detailed “fingerprint-like” information of the overall molecular composition of living bacterial cells in an extremely brief time span. Figure 7.2 shows a workflow as example. Raman spectra were collected from pure colonies that were grown under controlled conditions, here *Escherichia coli, Staphylococcus epidermidis* and *Bacillus subtilis* [6]. Compared to

Raman spectra of pure components in Figure 7.1, main bands are assigned to proteins and DNA. The scissoring and deformation vibrations of C–H bonds are found around 1,450 and 1,337 cm^{-1}, respectively. Vibrations of the peptide linkage of proteins are located around 1,660 cm^{-1} (amide I) and with less intensity around 1,242 cm^{-1} (amide III). The most prominent spectral contributions of aromatic amino acids are found at 850 and 1,003 cm^{-1}. Bands near 730, 750, 780 and 1,575 cm^{-1} are assigned to nucleotides and the band near 1,100 cm^{-1} to phosphate backbone of DNA. The band at 1,128 cm^{-1} is due to C–N and C–C stretching vibrations. Then, a database is compiled with Raman spectra representing biological variability – such as batch-to-batch variation or different incubation time – or mimicking the unknown history of individual bacterial cells obtained from patients or found in hospitals, clean room environments or food production lines. These data train a classification model. Details of chemometric data analysis are described in Chapter 4 of this book (Evaluation). Finally, Raman spectra of independent samples are used to validate the classification. In Figure 7.2, the accuracy of the model is compared with the morphological parameters such as the elongated shape of *B. subtilis*, the small size of *S. epidermidis* and single, double or triple spheres of *E. coli*.

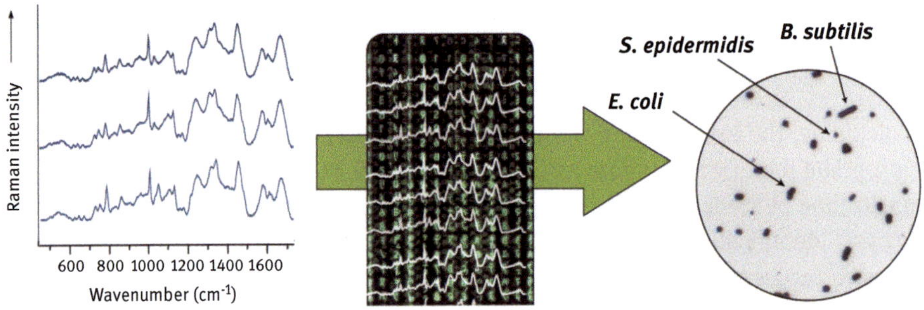

Figure 7.2: Workflow for micro-Raman spectroscopy of bacteria identification. Raman spectra from pure colonies (left) train a classification model (center) that is validated to assign independent samples of mixed bacteria (right). Adapted from Ref. [6].

An early report described the classification of 42 candida strains comprising 5 species by micro-Raman spectroscopy to demonstrate the feasibility of the technique for the rapid identification [7]. Raman spectroscopy was compared to 16s RNA sequencing for the identification of *Mycobacterium tuberculosis* complex strains and the most frequently found strains of nontuberculous mycobacteria (NTM) [8]. A total of 63 strains belonging to 8 distinct species were analyzed. The sensitivity of Raman spectroscopy for the identification for the identification of *Mycobacterium* species was 95.2 %. All *M. tuberculosis* strains were correctly identified (7 of 7; 100 %), as were 54 of 57 NTM strains (94 %). The differentiation between *M. tuberculosis* and NTM was invariably correct for all strains. A recent report showed almost 100 %

successful identification of staphylococci by Raman spectroscopy. Characteristic Raman spectra of 277 strains belonging to 16 species from a 24-h culture of each strain on a Mueller–Hinton agar plate were obtained [9]. Raman spectroscopy was used in the epidemiology of 525 isolates of methicillin-resistant *Staphylococcus aureus* (MRSA) of human- and animal-related clonal lineages [10]. When epidemiologically linked isolates from the 16 well-documented outbreaks were analyzed with pulsed-field gel electrophoresis (PFGE) as the reference standard, Raman spectroscopy correctly identified 97 % of cases that were indistinguishable from the index case. The authors concluded that Raman spectroscopy is a quick and reliable method of MRSA typing, which can be used in outbreak settings and it is comparable to PFGE, with the added advantage that PFGE non-typeable isolates can also be readily typed using the same sample preparation protocol. The Raman system SpectraCellRA (RiverD, The Netherland) was introduced as a fast easy-to-use and highly reproducible platform for bacterial typing and MRSA outbreak analysis [11].

7.2.2.2 Detection of infections

Urinary tract infections (UTI) belong to the most common infections. Micro-Raman spectroscopy in combination with chemometric classification was used as a culture-independent identification procedure to identify UTI causing microorganisms [12]. First, a Raman database of 11 relevant UTI bacterial species that were grown in sterile filtered urine was built. A classification model was trained and applied in a second step to identify infected urine samples of 10 patients and determine the predominant bacterial species. Another Raman setup was combined with dielectrophoresis to recover pathogens from the urinary tract [13]. The dielectrophoretic enrichment of bacteria allowed to obtain high-quality Raman spectra in dilute suspensions with an integration time of only 1 s. The setup demonstrated the Raman-based distinction of *E. coli* and *Enterococcus faecalis* – two bacterial strains that are commonly encountered in UTI.

Micro-Raman spectroscopy directly analyzed clinical relevant single bacterial cells from cerebrospinal fluid (CSF) during bacterial meningitis [14]. As proof of principle, a CSF sample obtained from a patient with meningococcal meningitis was investigated. As the CSF matrix did not mask the Raman spectrum of a bacterial cell, chemometric analysis of Raman spectra successfully identified *Neisseria meningitidis* cells from patients with bacterial meningitis.

Hospital acquired infections increase morbidity and mortality and constitute a high financial burden on health-care systems. Therefore, effective monitoring is important to detect potential outbreaks and sources of contamination. A microbial typing method was presented based on Raman spectroscopy which can be performed within minutes and is generally applicable as already written above [15]. 118 *S. aureus* isolates illustrated that the discriminatory power matches that of established genotyping techniques and that the concordance with the gold standard PFGE is high (95 %).

Enterococci have emerged as one of the leading causes of nosocomial infections worldwide. A comparative study uses phenotyping, genotyping and vibrational spectroscopy techniques for typing a collection of 18 *Enterococcus* strains comprising 6 different species [16]. Different laboratories using confocal micro-Raman spectroscopy, attenuated total reflection (ATR) infrared spectroscopy and Fourier-transform infrared (FTIR) microspectroscopy demonstrated the discriminative capacity and reproducibility of each technique.

An advanced Raman-based imaging approach provided false color images to specifically identify intracellular *S. aureus* and to localize them exactly in three dimensions within endothelial cells [17]. *S. aureus* is one of the most frequent human pathogens that can also act as a facultative intracellular pathogen causing infections that are extremely difficult to treat. The spectral information reveals that the intracellular bacteria are in the exponential growth phase with a reduced replication rate and biochemically different from extracellular bacteria proving their adaptation to the host's conditions.

7.2.2.3 Antibiotic resistance

After bacterial identification, the infectious ones require treatment with efficient antibiotics. A suitable target structure of many antibiotics e. g. of the group of the beta-lactams, penicillins and glycopeptides is the bacterial cell wall. Infections with multidrug-resistant microorganisms are an increasing threat to hospitalized patients. Raman spectroscopy was used to assess whether extended spectrum beta-lactamase (ESBL) producing *Klebsiella* species are transmitted between patients [18]. Isolates from 132 patients were analyzed and 17 clusters were identified with 17 primary and 56 secondary patients. Secondary patients were those identified with an isolate clonally related to the isolate of the primary patient. The authors concluded that Raman spectroscopy detected clonal outbreaks of ESBL-producing *Klebsiella* species in a hospital setting.

Vancomycin is an important glycopeptide antibiotic. During the last years, vancomycin resistances, especially among Enterococci, have risen. Raman spectroscopy was applied to characterize vancomycin–enterococci interactions over a time span of 90 min using a sensitive *E. faecalis* strain and two vancomycin concentrations above the minimal inhibitory concentration (MIC) [19]. Successful action of the drug on the pathogen could be observed already after 30 min interaction time.

Raman spectroscopy was able to detect the antimicrobial susceptibility of bacterial species isolated from a positive blood culture bottle within 5 h [20]. Full concordance with the reference methods VITEK®2 data and broth dilution was obtained for the antibiotic-susceptible strains, 68 % and 98 %, respectively, for the resistant strains.

Raman spectroscopy was used to profile the phenotypic response of *E. coli* to applied antibiotics [21]. *E. coli* cultures were subjected to three times the MICs of

15 different antibiotics with known mechanisms of action for 30 min before being analyzed by Raman spectroscopy. Chemometric analysis showed the ability of Raman spectroscopy to predict the functional class of an unknown antibiotic and to identify individual antibiotics that elicit similar phenotypic responses.

An integrated microfluidic device was presented in which bacteria from diluted suspensions were captured in well-defined regions using on-chip dielectrophoresis. Raman spectroscopy was used for further analysis to determine antibiotic susceptibilities of bacterial pathogens [22]. Ciprofloxacin-resistant *E. coli* was differentiated from sensitive *E. coli* with high accuracy within roughly 3 h total analysis time paving the way for future point-of-care devices.

7.2.2.4 Biofilms

Biofilms are communal living way of microorganisms in which bacterial cells are surrounded by extracellular polymeric substances. Most of the microorganisms live in biofilm form in the world. Since microorganisms are everywhere, understanding biofilm structure and composition is extremely crucial to make the world a better place for living not only for humans but also for other living beings. Raman spectroscopy was applied as a nondestructive technique in numerous papers to study biofilms.

Dehydrated biofilms of *Streptococcus mutans* and *Streptococcus sanguinis* were studied as model of dental plaque by confocal micro-Raman spectroscopy [23]. A prediction model based on principal component analysis and logistic regression was calibrated using pure biofilms of each species. When biofilms of the two species are partially mixed together, Raman-based identification was achieved within 2 μm of the boundaries between species with 97 % accuracy.

An interesting metabolic fingerprinting approach combined Raman spectroscopy with stable isotope probing to demonstrate the quantitative labeling and differentiation of *E. coli* cells [24]. *E. coli* cells were grown in minimal medium with fixed final concentrations of carbon and nitrogen supply, but with different ratios and combinations of 13C/12C glucose and 15N/14N ammonium chloride, as the sole carbon and nitrogen sources, respectively. The multivariate analysis investigation of Raman data illustrated unique clustering patterns resulting from specific spectral shifts upon the incorporation of different isotopes, which were directly correlated with the ratio of the isotopically labeled content of the medium.

7.3 Micro-Raman spectroscopy of cells

7.3.1 Experimental considerations

Single cells are very suitable objects for micro-Raman spectroscopy because of the high concentrations of biomolecules in their condensed volume. Protein concentration as high as 250 µg/µl and DNA and RNA concentrations in the range of 100 µg/µl have been reported [25]. These values depend on the cell type, the phase of the cell cycle and the location within the cell. Single eukaryotic cells are generally larger than prokaryotic bacteria that have been presented in the previous section. Therefore, a single Raman spectrum with a diffraction limited laser focus is not able to assess a single cell with its subcellular heterogeneity. One solution is to acquire micro-Raman images. However, collecting a Raman image from a $20 \times 20\,\mu m^2$ area with 0.5 µm step size with 1 s exposure time per spectrum takes ca. 30 min. This is not appropriate for high-throughput applications. Other solutions aim to acquire an average Raman spectrum per cell using an expanded laser focus or scanning a small laser over the cell during signal collection. The latter approach has been called integrated Raman acquisition in a recent paper [26]. It is well known that green laser excitation irreversibly damages single living cells. Cytochrome proteins that are involved in metabolic pathways within the cell organelles' mitochondria partly absorb this wavelength and their Raman spectra are resonantly enhanced for ca. 1 min. After this time, signals decrease due to cytochrome degradation. Near-infrared excitation does not harm single living cells even at high intensities and prolonged exposure times [27, 28]. Chemically fixed cells can be studied by all excitation wavelengths with the advantage of higher scattering intensities at shorter wavelengths. According to the scattering law, the intensities increase by $2^4 = 16$ if 400 nm instead of 800 nm is used for excitation.

Micro-Raman imaging was combined with independent but simultaneous optical microscopy such as phase measurements of cells [29, 30]. The resulting data provided information on how the light is retarded and/or scattered by molecules in the cell. The authors showed how the chemistry of a cell highlighted in the Raman information is related to the cell quantitative phase information, and they presented a technical overview of a multimodal system combining micro-Raman spectroscopy and quantitative phase microscopy. The dynamic measurements obtained from phase microscopy also enabled to monitor the cell morphology during the laser scanning of the Raman acquisition, making it possible to identify movements which may occur during the experiment.

Cells can be prepared in culture media under controlled conditions from well-defined cell lines. As the composition of the media might affect the Raman spectra of single cells, same media should be used if Raman spectroscopy is used for comparison of cell lines. Cells can also be prepared by extraction from body fluids such as blood, urine, saliva or CSF. These cells are called primary cells. Protocols were published to prepare biological materials for Raman spectroscopy that also include fixed cells [31].

7.3.2 Cytological applications of micro-Raman spectroscopy

Micro-Raman spectroscopy studies in the context of single cells have been reported for label-free classification, drug–cell interaction and cell metabolism. Raman spectra of single cells provide a specific fingerprint that can be used to determine cell types (e. g. normal vs. tumor), cell stages (e. g. normal vs. apoptotic) or proliferation states. Promising applications in cytology are the identification of tumor cells that circulate (CTCs) in blood of cancer patients, screening of abnormal cells in cervix uteri and bladder cancer cells in urine. Examples are given next.

7.3.2.1 Classification of single cells

Because CTCs are extremely rare (ca. 1 in 1 million leukocytes), model systems were studied that use leukocytes from control patients, leukemia cells (OCI-AML3) and breast tumor cells (MCF-7 and BT-20). Raman-based classification rates of these cells were reported between 99.7 % and 92 % for images of cells after drying [32], for single spectra after optical tweezing in aqueous buffer [33], for spectra after optical trapping in microfluidic chips made of glass [34] and quartz [35]. A typical workflow which was applied in Ref. [32] is shown in Figure 7.3 to identify CTCs by micro-Raman spectroscopy. Raman images of single cells were collected at small step size. The mean spectra were used to train a classification model. A cell mixture was prepared, a Raman image at larger step size was collected from a region of interest and the data were subjected to the classification model. The color-coded classification results were validated by immunofluorescent staining. The comparison reveals that tumor cells were correctly identified. Even the breast tumor cells MCF7 and BT20 were distinguished by Raman spectroscopy which was not the case for immunofluorescence because the

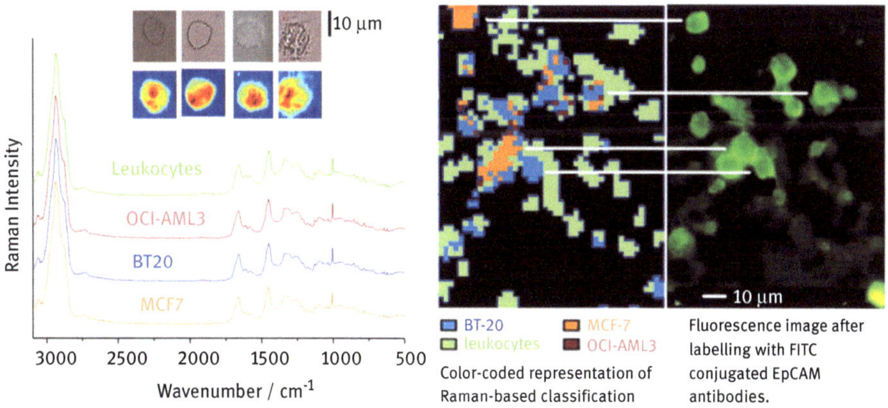

Figure 7.3: Workflow of micro-Raman spectroscopy of cell identification. High-resolution Raman images of single cells are collected and mean spectra are calculated that train a classification model. The model assigns each spectrum of a low resolution Raman image of a mixed cell population to a cell type and the color-coded classification results were validated by immunofluorescent staining. Adapted from Ref. [32].

EpCAM antigen was present in both tumor cells. Furthermore, leukocytes were identified by Raman spectroscopy whereas they could not be detected by immunofluorescence due to the absence of the antigen. Although this workflow was accurate, the throughput was too low for clinical translation. A high-throughput screening Raman approach was recently introduced [36] which combines a custom-built microscope and an automated workflow with an image-based cell localization. Instead of collecting a Raman image of the whole slide which also contains areas without cells as evident in Figure 7.3, the positions of cells were determined and only a single spectrum of each cell was collected. The high-throughput screening approach was applied to obtain Raman spectroscopic blood cell counts. More than 100,000 Raman spectra were collected from 3 leukocyte subtypes, namely neutrophils, monocytes and lymphocytes to train a classification model. Then, this model was applied to mixed leukocyte populations of two healthy volunteers and three patients. It was found that the Raman-based results agreed well with machine counting as reference.

High-grade squamous intraepithelial cells (HSIL) were detected by investigating diseases associated changes in morphologically normal appearing intermediate and superficial cells using Raman spectroscopy [37]. Raman spectra were measured from intermediate and superficial cells from negative and HSIL cytology ThinPrep specimens. Partial least squares discriminant analysis (PLS-DA) provided sensitivities of 95.5 %, 95.2 % and 96.1 % and specificities of 92.7 %, 94.7 % and 93.5 % for HSIL diagnosis based on the dataset obtained from intermediate, superficial and mixed intermediate/superficial cells, respectively. Similarly, high-grade urothelial carcinoma in urine was noninvasively diagnosed by Raman spectral imaging [38]. Coherent anti-Stokes Raman scattering (CARS) imaging of urine sediments was used in the first step for fast preselection of urothelial cells where high-grade urothelial cancer cells were characterized by a large nucleus to cytoplasm ratio. In the second step, Raman spectral imaging of urothelial cells was performed. A supervised classifier was implemented to automatically differentiate normal and cancerous urothelial cells with 100 % accuracy. Raman marker bands showed a decrease in the level of glycogen and an increase in the levels of fatty acids in cancer cells as compared to controls.

Two liver cancer cell lines Hep-G2 and SK-Hep1 were distinguished by Raman imaging [39]. Support vector machine-based classification predicted independent cells and cell cycle phases. The datasets were segmented by cluster analysis. The classifier was more accurate if spectra of the cytoplasm were used as input instead of average spectra or spectra of cell nuclei. The spectral variations were assigned to altered unsaturated fatty acids content. The same variations were found for the discrimination of isogenic cancer cells by Raman imaging [40]. Here, classification was achieved by PLS-DA. Raman spectroscopy distinguished myeloblasts, promyelocytes, abnormal promyelocytes and erythroblasts which have to be counted for a correct diagnosis and morphological classification of acute myeloid leukemia and myelodysplastic syndrome [41]. Cells from bone marrow samples of patients were analyzed by Raman imaging. Leave-one-out cross-validation of a classification model based on linear discriminant analysis (LDA) of average Raman spectra gave accuracies between 95 % and 100 % to identify each cell type. Raman spectroscopy characterized and differentiated two breast cancer and one normal breast cell line [42]. Principal component analysis identified variations for accurate and reliable separation of the three cell lines. A LDA model classified the cell lines with 100 % sensitivity and 91 % specificity.

A multimodal microscopy platform combined micro-Raman spectroscopy and multispectral imaging to distinguish lymphocytes from 24 leukemic patients and 11 healthy individuals [43]. Blood smears were prepared on microscopy slides, 90 cells per blood sample were identified by optical microscopy and Raman spectra were acquired on cell nuclei. Raman data were classified using a support vector machine algorithm that distinguished lymphocytes from other nucleated blood components with 99.6 % sensibility and 98.8 % specificity.

7.3.2.2 Cell metabolism

In an early study, Raman images from normal and oxidative stressed lung fibroblast cells were collected at 1 µm step size [44]. Oxidative stress was induced by treatment with glyoxal for 24 h. Cells were fixed with formalin and transferred to phosphate buffered saline before micro-Raman spectroscopy with a 60× water immersion objective lens. The datasets were segmented by cluster analysis which visualized the cell morphology in cluster membership maps and the spectral signatures of the relevant organelles in cluster mean spectra. Among the most evident changes were that the stressed cells have smaller cell nuclei and lower DNA content which is consistent with inhibition of DNA replication, higher DNA condensation and shrinkage of a pyknotic cell nucleus.

Raman spectroscopy was used as a label-free method to investigate the biochemical changes occurring in macrophages during the first few hours of hemozoin uptake [45]. The so-called malaria pigment hemozoin is engulfed by phagocytic cells, such as macrophages, during malaria infection. The spatial distribution of hemozoin in Raman images was found to be inhomogeneous, and its presence largely excluded that of proteins and lipids, demonstrating that the cells were not able to break down the biocrystals on the studied time scale.

The cell cycle phase was determined in human embryonic stem cells by Raman imaging [46]. The nucleic acid band at 783 cm^{-1} was integrated across individual cell nuclei and corrected for RNA contributions. The measured intensities reflected DNA content and exhibited a profile analogous to that obtained during flow cytometry using nuclear stains. Raman imaging at 633 nm excitation investigated living cells during cytokinesis, the climax of the cell cycle [47]. Ten Raman images of dividing and nondividing human colon cancer cells were analyzed by multivariate curve resolution. Intrinsic Raman spectra and intracellular distributions of major biomolecular components were derived and an unusual autofluorescent lipid component was discovered that appeared predominantly in the vicinity of the cleavage furrow during cytokinesis.

If the small molecules of interest do not contain groups that would allow for a discrimination against cellular background signals, "labeling" of the molecule by isotope substitution, or by other small groups, e. g. alkynes, provides a stable Raman signal. This approach to exploit such small reporters that possess Raman signals without overlap with naturally existing biomolecules in cells is also called bioorthogonal Raman imaging. The uptake of lipids by macrophages was studied by Raman imaging [48]. Here, the polyunsaturated arachidonic acid was deuterated to distinguish this fatty acid from other, non-deuterated lipids. Arachidonic acid was found to be stored in droplets, but foam cell formation is less pronounced than with palmitic acid and oleic acid studied earlier [49]. Stable isotope labeling by amino acids in cell culture was combined with Raman spectroscopy to selectively monitor the incorporation of deuterium-labeled phenylalanine (L-Phe (D8)) from the epithelial cell line ARPE-19 into proteins of tachyzoites of the parasite *Toxoplasma gondii* [50]. L-Phe(D8) from the host cell completely replaces the L-Phe within tachyzoites 7–9 h after infection. A quantitative model based on Raman spectra allowed an estimation of the exchange rate of Phe as $0.5–1.6 \times 10^4$ molecules/s.

7.3.2.3 Drug–cell interaction

The cellular responses to a drug were studied without marker by micro-Raman imaging. This was demonstrated by measuring the effect of the epidermal growth factor receptor (EGFR) inhibitor panitumumab on cells lines [51]. The Raman study detected large panitumumab-induced differences in cells expressing wild-type K-ras, but not in cells with oncogenic K-ras mutations. Other works monitored the effects of the anticancer drug docetaxel on the morphology and biochemistry of living colon cancer cells [52], the anticancer drug paclitaxel in MCF-7 breast cancer cells [53, 54], the chemotherapeutic drug doxorubicin on lymphocytes [55] and the antitumor drugs gemcitabine on living non-small lung cancer cells Calu-1 [56] and cisplatin on nasopharyngeal carcinoma cells [57] by micro-Raman imaging.

Micro-Raman spectroscopy was applied to show the spatial distributions of the drug erlotinib within the cell [58]. Erlotinib was identified by its alkyne group as bioorthogonal label introduced in Section 3.2.2. The Raman images showed that the drug is clustered at the EGFR protein at the membrane and induces receptor internalization. The changes within the Raman spectrum of erlotinib measured in cells as compared to the free-erlotinib spectrum indicated that erlotinib is metabolized within cells to its demethylated derivative.

7.4 Micro-Raman spectroscopy of tissues

7.4.1 Experimental considerations

Advantages of micro-Raman spectroscopy of tissue include that no or minimal sample preparation is required and data acquisition is nondestructive which enables also *in vivo* studies. Samples can be prepared as thin cryosections or formalin-fixed and paraffin-embedded tissue sections that can be compared with histopathology findings after staining. Investigations of fresh biopsies, organs and living organisms enable a plethora of Raman-based applications. Early reports in the literature regarding the utility of micro-Raman spectroscopy to tissues were based on acquisition of spectral data only at single points which required an a priori knowledge of the location or a preselection of the probed position. Since the inhomogeneous nature of tissue on the microscopic level was not considered in these early studies, an accurate correlation between the histopathology of the sampled area and the corresponding spectra was only possible for homogenous tissue on the macroscopic level. Considerable progress was made when high-throughput and more sensitive instruments became available for micro-Raman imaging. They enable to microscopically collect larger number of spectra from larger sample populations in less time, improving statistical significance and spatial specificity. Micro-Raman images with cellular resolution require a diffraction-limited laser focus diameter and step size near 2 µm. Then, cell nuclei can be resolved. However, this has only been demonstrated in small regions of interest below $200 \times 200\,\mu m^2$. For regions of interest above $2 \times 2\,mm^2$, more than 1 million Raman need to be collected which would be extremely time consuming in sequential raster scan acquisition mode. One option is to expand the laser focus diameter and increase the step size which harbors the risk to miss details in such an undersampled, coarse overview. Another option is parallel data acquisition by line scanning or global imaging. Finally, combination with other modalities e. g. autofluorescence or OCT enables rapid prescreening and selecting smaller regions-of-interest for micro-Raman imaging. An experimentally complex alternative is coherent Raman scattering (CRS) using high-energy short-pulse lasers such as CARS and stimulated Raman scattering. CRS can increase the intensities of defined vibrational bands by six orders of magnitude which enables rapid acquisition in the range of microseconds per pixel and video-rate image registration at cellular resolution with laser scanning microscopes. A detailed description is not within the scope of this chapter and can found elsewhere [59]. Examples are presented for the other options mentioned above.

A challenge for Raman spectroscopy of most tissue samples is an autofluorescence background that is generated by the excitation laser and overlaps with the weaker Raman bands. To minimize this background, near infrared lasers are usually used as excitation source. The frequently used wavelength of 785 nm represents a compromise between low fluorescence, reasonable Raman scattering intensities and quantum efficiency of charge coupled device (CCD) detectors. However, even with 785 nm excitation the fluorescence background from some samples still disturbs. Data processing tools have been developed to correct fluorescence background variations e. g. subtracting a polynomial function. Data collection approaches to suppress fluorescence in Raman spectroscopy were reviewed [60] and include time gating to separate the rapid Raman scattering from the slower fluorescence, shifted subtracted Raman spectroscopy, shifted excitation Raman difference spectroscopy and multi-excitation Raman spectroscopy. The principle of these approaches is that the fluorescence emission does not significantly vary upon small wavelength shifts, whereas the Raman bands shift as a function of the excitation wavelength and Raman signatures can be obtained by calculating differences.

7.4.2 Applications of micro-Raman imaging

Micro-Raman spectroscopy to detect, diagnose and delineate cancer tissues is a promising complementary tool in histopathology [61]. Whereas tissue types can be identified from single Raman spectra, Raman imaging is important for three reasons. First, Raman images of unstained tissue section and subsequent comparison with the golden standard, usually a hematoxylin and eosin (H&E) stained tissue section, are the method of choice to develop and validate Raman-based classification models. After validation, the model can be applied to single spectra. Second, Raman images allow accurate determination of margins between normal and non-normal tissue, e. g. inflammation, dysplasia, tumor or necrosis. Third, Raman images resolve morphological details without staining which give complementary information to the molecular information of the spectra.

7.4.2.1 Lateral resolution in micro-Raman images

Figure 7.4 shows Raman images from an unstained colon tissue cryosection that were collected at step size of 62 μm (b), 10 μm (c) and 2.5 μm (d). The image sizes range from $4.4 \times 4.4\,mm^2$ (72 × 72 pixel, b), $520 \times 510\,\mu m^2$ (53 × 52 pixel, c) to $147 \times 147\,\mu m^2$ (59 × 59 pixel, d). The box in (a) indicates the region-of-interest in (c), and the box in (c) the region-of-interest in (d). Cluster membership map (b) correlates with the main tissue types serosa (red), longitudinal (olive) and circular (yellow) muscle layers, submucosa (brown), mucosa (blue) and mucus (magenta). Ganglia are evident between longitudinal and circular muscle layers that are not resolved in Raman image (b), but in (c) at smaller step size and higher lateral resolution (gray cluster). The scattered appearance is due to noise in Raman spectra. More details even at subcellular level are resolved for a single ganglion in (d). In Figure 7.4, the photomicrograph (a) gave sufficient morphological details to select regions-of-interest. Furthermore, the tissue section was prepared on an UV-grade calciumfluoride slide

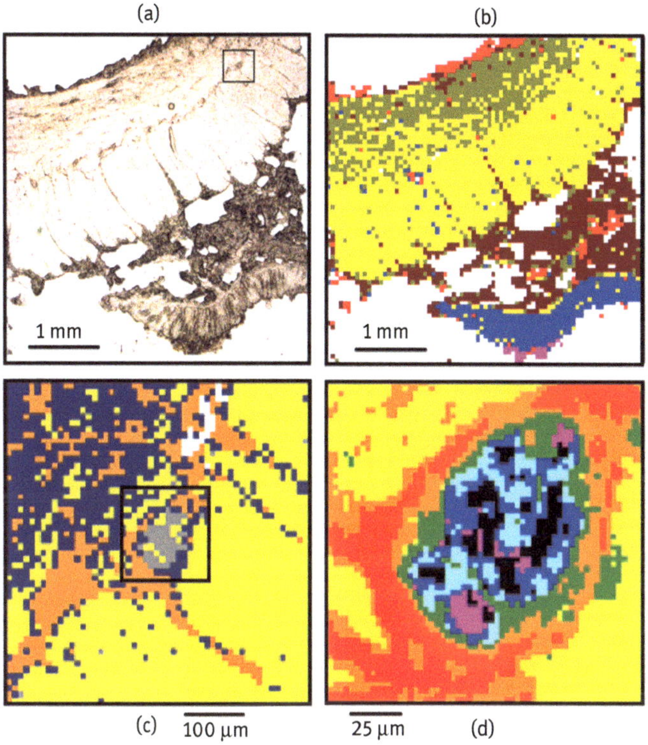

Figure 7.4: Raman images from an unstained colon tissue cryosection (a) that were collected at step size of 62 µm (b), 10 µm (c) and 2.5 µm (d). The image sizes range from 4.4 × 4.4 mm^2 (72 × 72 pixel, b), 520 × 510 μm^2 (53 × 52 pixel, c) to 147 × 147 μm^2 (59 × 59 pixel, d). The box in (a) indicates the region-of-interest in (c), and the box in (c) the region-of-interest in (d). The color codes represent the segmentation by cluster analysis. Adapted from Ref. [62].

which gave a low Raman background and enabled FTIR imaging for rapid prescreening. FTIR images, IR and Raman spectra and more details were published [62].

7.4.2.2 Raman-based detection of pathologies

Micro-Raman images of six dried brain tumor tissue sections of different malignancy were collected and processed by vertex component analysis (VCA) instead of cluster analysis [63]. VCA decomposes Raman images into spectral signatures and their abundance of the most dissimilar components. Cell density and cell nuclei obtained from VCA abundance plots correlated well with malignancy obtained from H&E stained tissue sections. Furthermore, Raman spectra of high-grade tumor cell nuclei were found to contain more intense spectral contributions of nucleic acids than those of low-grade tumor cell nuclei. This approach was transferred to non-dried tissue sections in aqueous medium from which Raman images were collected with a water immersion objective lens. Here, dehydration-induced denaturation of proteins, nucleic acids and lipids did not occur which gave rise to more intense signals from cell nuclei and avoid formation of lipid/cholesterol microcrystals [64]. The hyperspectral unmixing algorithms VCA and N-FINDR calculated seven components that were assigned to cholesterol, cholesterol ester, DNA, carotene, proteins, lipids and buffer and correlated well with the Raman spectra of reference materials. Another paper defined the most abundant and most dissimilar components by hyperspectral unmixing and used these components as input for nonlinear least square fitting to determine protein, nucleic acid and lipid content and the lipid to protein ratio [65]. Image analysis revealed the average nuclei size, the number of nuclei and the fraction of nuclei area. Except for the protein content which was found to be almost constant, all chemical and morphological parameters correlated with the malignancy of glioma brain tumors. The primary tumor of brain metastases could be identified by Raman imaging and hierarchical support vector machines [66]. The first level discrimination distinguished normal brain, necrosis and tumor tissue. The second level of discrimination used tumor tissue Raman spectra as input and determined the primary tumor of bladder carcinoma, mamma carcinoma, colon carcinoma, prostate carcinoma and renal cell carcinoma in 22 cyrosectioned specimens with accuracy better than 99 %. In contrast to the work presented before, Gajjar et al. [67] prepared tissue sections from formalin-fixed paraffin-embedded tissue blocks of normal brain and different brain tumors. Before ATR-FTIR spectroscopy and Raman spectroscopy, samples were dewaxed, mounted on low-E slides and desiccated. This preparation removes virtually all hydrophobic molecules including lipids. Nevertheless, normal and different tumor subtypes were segregated.

Raman imaging was applied to analyze normal, benign and cancerous breast tissue of 146 patients [68]. The results provided evidence that carotenoid, lipid and fatty acid composition of breast cancer differed significantly from that of surrounding noncancerous breast tissue. Using Raman imaging with 532 nm excitation, spectral histopathology of colon cancer tissue sections was shown to achieve 10 times higher spatial resolution as compared to FTIR imaging. In addition to visualization of erythrocytes and lymphocytes in Raman images, autofluorescence was found to spatially overlap with the fluorescence of antibodies against p53 [69]. Sentinel lymph node biopsy has become the standard surgical procedure for the sampling of axillary lymph nodes in breast cancer. Fifty-nine axillary lymph nodes, 43 negative and 16 positive from 58 patients undergoing breast surgery, were mapped using micro-Raman spectroscopy [70]. These maps were then used to model the effect to use a Raman spectroscopic probe by selecting 5 and 10 probe points across the mapped images and evaluating the impact on disease detection. Results demonstrated sensitivities of up to 81 % and specificities of up to 97 % when differentiating between positive and negative lymph nodes, dependent on the number of probe points included. Micro-Raman imaging characterized inflammatory bowel diseases in colon tissue sections [71].

First, support vector machine classification was applied to highlight the tissue morphology. In a second step, the biochemical tissue composition was studied by analyzing the epithelium Raman spectra in sections of healthy control subjects, subjects with Crohn's disease (CD), and subjects with ulcerative colitis (UC). A support vector machine classifier distinguished healthy control patients ($n = 11$), patients with CD ($n = 14$), and patients with UC ($n = 13$) with an accuracy of 98.90 %. Raman spectroscopy enabled label-free detection of peripheral nerve tissue against adjacent tissue [72]. Raman spectra of myelinated nerve, unmyelinated nerve, fibrous connective tissue, skeletal muscle, tunica media of blood vessel and adipose tissue of Wistar rats were analyzed, and Raman images of the tissue distribution were constructed using the map of the ordinary least squares regression estimates. Nerve tissue was found to exhibit a specific Raman spectrum arising from axon or myelin sheath. This method was also applied to unfixed tissue sections of human periprostatic samples excised from prostatic cancer patients. The potential of Raman spectroscopy for oral cancer detection in surgical margins was investigated [73]. One hundred and twenty-seven Raman images were collected from tissue sections of 10 patients. Spectra annotated as oral cavity squamous cell carcinoma (OCSCC) or surrounding healthy tissue structures were used as input for LDA. The LDA model distinguished OCSCC spectra from spectra of adipose tissue, nerve, muscle, gland connective tissue and squamous epithelium. Structures that were most often confused with OCSCC were tissues close to OCSCC. This result demonstrated how well Raman spectroscopy enabled discrimination between OCSCC and surrounding healthy tissue structures. Renal oncocytoma and chromophobe renal cell carcinoma are two kidney tumors which are sometimes confused using conventional histopathological methods. In an investigation based on 88 patients, a partial least squares discriminant analysis model generated from Raman images distinguished between both carcinomas with a performance of 86 % sensitivity and 81 % specificity [74].

7.4.2.3 Combination with other modalities

Multiple modalities are combined in efforts to improve the performance of optical methods for disease detection. Current interests in that direction are developments of dual-modal instruments or applications of complementary methods combining the biochemical specificity of Raman spectroscopy with the sensitivity and rapid screening capabilities of other label-free high-resolution optical imaging like OCT, FTIR imaging, fluorescence and phase contrast microscopy.

Biochemical markers of atherosclerotic plaque from mice were determined by Raman, infrared and AFM imaging [75]. Label-free imaging of cholesterol, cholesteryl esters, remodeled media, heme, internal elastic lamina, fibrous cap and calcification provided additional knowledge of plaque features. FTIR and Raman images were compared in tissue sections of rabbit atherosclerotic plaque depositions [76]. Raman imaging was found to differentiate more biochemical components, namely proteins, cholesterol ester, cholesterol and triglycerides and also resolved more structural details. FTIR imaging using 64 × 64 MCT focal plane array detector enabled faster data acquisition which is particularly important for full arterial cross sections. Using a non-destructive approach involving Raman and atomic force microscopy imaging, the presence of clustered lipid rafts in endothelium of the aorta of the db/db mice that represent a reliable murine model of type 2 diabetes has been demonstrated [77]. The raft clusters in the aorta of diabetic mice were shown to occupy a considerably larger (about 10-fold) area of endothelial cell surface as compared to the control.

An integrated optical technique was demonstrated for basal cell carcinoma of skin based on autofluorescence imaging and Raman scattering [78]. Automated segmentation of autofluorescence images was used to select and prioritize sparse sampling points for Raman spectroscopy. This automated sampling strategy yielded classification models with 100 % sensitivity and 92 % specificity and allowed objective diagnosis faster than frozen section histopathology and faster than FTIR or Raman imaging alone. The approach was transferred to the diagnosis of mammary ductal carcinoma [79] and basal cell carcinoma in skin tissue samples [79]. Raster scanning micro-Raman spectroscopy of tissue sections is slow with typically 10,000 spectra/mm^2 which is equivalent to 5 h/mm^2. Selective sampling based on integrated autofluorescence and Raman spectroscopy reduced the number of Raman spectra to 20 spectra/mm^2 which is equivalent to an acquisition time of 15 min for 5 × 5 mm^2.

A clinical instrument was developed to combine Raman spectroscopy and OCT for morphological and biochemical characterization of skin cancer [80]. The instrument utilized independent 785 nm Raman spectroscopy and 1,310 nm OCT system backbones and enabled sequential acquisition of co-registered OCT and Raman datasets. Ashok et al. reported a multimodal optical approach using Raman spectroscopy and OCT in tandem to discriminate between colonic adenocarcinoma and normal colon [81]. The chemical information derived from Raman spectroscopy was combined with texture parameters extracted from OCT images which increased both sensitivity and specificity to 94 %.

A workflow was presented to combine Raman spectroscopic and MALDI mass spectrometric imaging for tissue studies [82]. The authors demonstrated for a larynx carcinoma sample that tissue types and different metabolic states could be extracted from Raman spectra. MALDI spectra yielded a better characterization of epithelial differentiation and understanding of ongoing dysplastic alterations.

7.5 Outlook

An overview has been given on micro-Raman spectroscopy in medicine. Examples were described in the context of bacteria, cells and tissues. Further developments are required to clinical translation. Instead of research grade microscopic systems, dedicated Raman systems should offer a high degree of automation. The workflow should follow standard operating procedures for sample preparation, and then the user is guided through the acquisition routine. This routine should also include instrument calibration to compensate for possible day-to-day fluctuations. As variations in Raman spectra of biomedical samples are often small, such fluctuations have a negative effect on the robustness and accuracy of classification models. Larger patient cohorts involving different clinical users, laboratories and instruments should validate the results and demonstrate the value of micro-Raman spectroscopy in future medicine. Prospects to solve unmet needs exist in point-of-care applications and personalized treatment of diseases.

References

[1] Krafft C, von Eggeling F, Guntinas-Lichius O, Hartmann A, Waldner MJ, Neurath MF, et al. Ex-vivo and in vivo optical molecular pathology. J Biophotonics. 2018;11:e201700236.

[2] De Gelder J, De Gussem K, Vandenabeele P, Moens L. Reference database of Raman spectra of biological molecules. J Raman Spectrosc. 2007;38:1133–47.

[3] Maquelin K, Kirschner C, Choo-Smith LP, van den Braak N, Endtz HP, Naumann D, et al. Identification of medically relevant microorganisms by vibrational spectroscopy. J Microbiol Meth. 2002;51:255–71.

[4] Pahlow S, Meisel S, Cialla-May D, Weber K, Rösch P, Popp J. Isolation and identification of bacteria by means of Raman spectroscopy. Adv Drug Del Rev. 2015;89:105–20.

[5] Ramoji A, Galler K, Glaser U, Henkel T, Mayer G, Dellith J, et al. Characterization of different substrates for Raman spectroscopic imaging of eukaryotic cells. J Raman Spectrosc. 2016;47:773–86.

[6] Pahlow S, Kloß S, Blättel V, Kirsch K, Hübner U, Cialla D, et al. Isolation and enrichment of pathogens with a surface-modified aluminium chip for Raman spectroscopic applications. Chem Phys Chem. 2013;14:3600–5.

[7] Maquelin K, Choo-Smith L-P, Endtz HP, Bruining H, Puppels G. Rapid identification of Candida species by confocal Raman microspectroscopy. J Clin Microbiol. 2002;40:594–600.

[8] Buijtels PC, Willemse-Erix H, Petit P, Endtz HP, Puppels GJ, Verbrugh HA, et al. Rapid identification of mycobacteria by Raman spectroscopy. J Clin Microbiol. 2008;46:961–5.

[9] Rebrošová K, Šiler M, Samek O, Ruzicka F, Bernatova S, Hola V., et al. Rapid identification of staphylococci by Raman spectroscopy. Sci Rep. 2017;7:14846.

[10] Wulf M, Willemse-Erix D, Verduin C, Puppels G, van Belkum A, Maquelin K. The use of Raman spectroscopy in the epidemiology of methicillin-resistant Staphylococcus aureus of human- and animal-related clonal lineages. Clin Microbiol Infect. 2012;18:147–52.

[11] Te Witt R, Vaessen N, Melles D, Lekkerkerk WS, van der Zwaan EA, Zandijk WH, et al. Good performance of the SpectraCellRA system for typing of methicillin-resistant staphylococcus aureus isolates. J Clin Microbiol. 2013;51:1434–8.

[12] Kloss S, Kampe B, Sachse S, Rösch P, Straube E, Pfister W, Kiehntopf M, Popp J. Culture independent Raman spectroscopic identification of urinary tract infection pathogens: a proof of principle study. Anal Chem. 2013;85:9610–6.

[13] Schröder U-C, Ramoji A, Glaser U, Sachse S, Leiterer C, Csaki A, et al. Combined dielectrophoresis-Raman setup for the classification of pathogens recovered from the urinary tract. Anal Chem. 2013;85:10717–24.

[14] Harz M, Kiehntopf M, Stöckel S, Rösch P, Straube E, Deufel T, et al. Direct analysis of clinical relevant single bacterial cells from cerebrospinal fluid during bacterial meningitis by means of micro-Raman spectroscopy. J Biophotonics. 2009;2:70–80.
[15] Willemse-Erix DF, Scholtes-Timmerman MJ, Jachtenberg J-W, van Leeuwen WB, Horst-Kreft D, Bakker Schut TC, et al. Optical fingerprinting in bacterial epidemiology: Raman spectroscopy as a real-time typing method. J Clin Microbiol. 2009;47:652–9.
[16] Kirschner C, Maquelin K, Pina P, Ngo Thi NA, Choo-Smith LP, Sockalingum GD, et al. Classification and identification of enterococci: a comparative phenotypic, genotypic, and vibrational spectroscopic study. J Clin Microbiol. 2001;39:1763–70.
[17] Große C, Bergner N, Dellith J, Heller R, Bauer M, Mellmann A, et al. Label-free imaging and spectroscopic analysis of intracellular bacterial infections. Anal Chem. 2015;87:2137–42.
[18] Voor in 'T Holt AF, Severin JA, Goessens WH, Te Witt R, Vos MC. Instant typing is essential to detect transmission of extended-spectrum beta-lactamase-producing Klebsiella species. PLoS One. 2015;10:e0136135.
[19] Assmann C, Kirchhoff J, Beleites C, Hey J, Kostudis S, Pfister W, et al. Identification of vancomycin interaction with Enterococcus faecalis within 30 min of interaction time using Raman spectroscopy. Anal Bioanal Chem. 2015;407:8343–52.
[20] Dekter H, Orelio C, Morsink M, Tektas S, Vis B, Te Witt R, et al. Antimicrobial susceptibility testing of Gram-positive and-negative bacterial isolates directly from spiked blood culture media with Raman spectroscopy. Euro J Clin Microbiol Infectious Diseas. 2017;36:81–9.
[21] Athamneh A, Alajlouni R, Wallace R, Seleem M, Senger R. Phenotypic profiling of antibiotic response signatures in Escherichia coli using Raman spectroscopy. Antimicrob Agents Chemother. 2014;58:1302–14.
[22] Schröder UC, Kirchhoff J, Hübner U, Mayer G, Glaser U, Henkel T, et al. On-chip spectroscopic assessment of microbial susceptibility to antibiotics within 3.5 hours. J Biophotonics. 2017;10:1547–57.
[23] Beier BD, Quivey RG, Berger AJ. Identification of different bacterial species in biofilms using confocal Raman microscopy. J Biomed Opt. 2010;15:066001.
[24] Muhamadali H, Chisanga M, Subaihi A, Goodacre R. Combining Raman and FT-IR spectroscopy with quantitative isotopic labeling for differentiation of E. coli cells at community and single cell levels. Anal Chem. 2015;87:4578–86.
[25] Hildebrandt ER, Cozzarelli NR. Comparison of recombination in vitro and in E. coli cells: measure of the effective concentration of DNA in vivo. Cell. 1995;81:331–40.
[26] Schie IW, Kiselev R, Krafft C, Popp J. Rapid acquisition of mean Raman spectra of eukaryotic cells for a robust single cell classification. Analyst. 2016;141:6387–95.
[27] Puppels GJ, Olminkhof JH, Segers-Nolten GM, Otto C, De Mul FF, Greve J. Laser irradiation and Raman spectroscopy of single living cells and chromosomes: sample degradation occurs with 514.5 nm but not with 660 nm laser light. Exp Cell Res. 1991;195:361–7.
[28] Notingher I, Verrier S, Romanska H, Bishop AE, Polak JM, Hench LL. In situ characterisation of living cells by Raman spectroscopy. Spectros (Amsterdam, Netherlands). 2002;15:43.
[29] Pavillon N, Hobro AJ, Smith NI. Cell optical density and molecular composition revealed by simultaneous multimodal label-free imaging. Biophys J. 2013;105:1123–32.
[30] Pavillon N, Smith NI. Implementation of simultaneous quantitative phase with Raman imaging. EPJ Tech Instrum. 2015;2:5.
[31] Butler HJ, Ashton L, Bird B, Cinque G, Curtis K, Dorney J, et al. Using Raman spectroscopy to characterize biological materials. Nat Protoc. 2016;11:664.
[32] Neugebauer U, Clement JH, Bocklitz T, Krafft C, Popp J. Identification and differentiation of single cells from peripheral blood by Raman spectroscopic imaging. J Biophotonics. 2010;3:579–87.

[33] Neugebauer U, Bocklitz T, Clement JH, Krafft C, Popp J. Towards detection and identification of circulating tumour cells using Raman spectroscopy. Analyst. 2010;135:3178–82.
[34] Dochow S, Krafft C, Neugebauer U, Bocklitz T, Henkel T, Mayer G, et al. Tumour cell identification by means of Raman spectroscopy in combination with optical traps and microfluidic environments. Lab Chip. 2011;11:1484–90.
[35] Dochow S, Beleites C, Henkel T, Mayer G, Albert G, Clement J, et al. Quartz microfluidic chip for tumour cell identification by Raman spectroscopy in combination with optical traps. Anal Bioanal Chem. 2013;405:2743–6.
[36] Schie IW, Rüger J, Mondol AS, Ramoji A, Neugebauer U, Krafft C, et al. High-throughput screening Raman spectroscopy platform for label-free cellomics. Anal Chem. 2018;90:2023–30.
[37] Duraipandian S, Traynor D, Kearney P, Martin C, O'Leary JJ, Lyng FM. Raman spectroscopic detection of high-grade cervical cytology: using morphologically normal appearing cells. Sci Rep. 2018;8:15048.
[38] Yosef HK, Krauß SD, Lechtonen T, Jütte H, Tannapfel A, Käfferlein HU, et al. Noninvasive diagnosis of high-grade urothelial carcinoma in urine by Raman spectral imaging. Anal Chem. 2017;89:6893–9.
[39] Tolstik T, Marquardt C, Matthäus C, Bergner N, Bielecki C, Krafft C, et al. Discrimination and classification of liver cancer cells and proliferation states by Raman spectroscopic imaging. Analyst. 2014;139:6036–43.
[40] Hedegaard M, Krafft C, Ditzel HJ, Johansen LE, Hassing S, Popp J. Discriminating isogenic cancer cells and identifying altered unsaturated fatty acid content as associated with metastasis status, using k-means clustering and partial least squares-discriminant analysis of Raman maps. Anal Chem. 2010;82:2797–802.
[41] Vanna R, Ronchi P, Lenferink AT, Tresoldi C, Morasso C, Mehn D, et al. Label-free imaging and identification of typical cells of acute myeloid leukaemia and myelodysplastic syndrome by Raman microspectroscopy. Analyst. 2015;140:1054–64.
[42] Talari AC, Evans CA, Holen I, Coleman RE, Rehman IU. Raman spectroscopic analysis differentiates between breast cancer cell lines. J Raman Spectrosc. 2015;46:421–7.
[43] Klossa J, Daliphard S, Troussard X, Vielh P, Manfait M, Angulo J, et al. Using biophotonics techniques to retrieve prognostic intracellular signatures. Irbm. 2011;32:72–5.
[44] Krafft C, Knetschke T, Funk RH, Salzer R. Studies on stress-induced changes at the subcellular level by Raman microspectroscopic mapping. Anal Chem. 2006;78:4424–9.
[45] Hobro AJ, Pavillon N, Fujita K, Ozkan M, Coban C, Smith NI. Label-free Raman imaging of the macrophage response to the malaria pigment hemozoin. Analyst. 2015;140:2350–9.
[46] Konorov SO, Schulze HG, Piret JM, Blades MW, Turner RF. Label-free determination of the cell cycle phase in human embryonic stem cells by Raman microspectroscopy. Anal Chem. 2013;85:8996–9002.
[47] Hsu J-F, Hsieh P-Y, Hsu H-Y, Shigeto S. When cells divide: label-free multimodal spectral imaging for exploratory molecular investigation of living cells during cytokinesis. Sci Rep. 2015;5:17541.
[48] Stiebing C, Matthaeus C, Krafft C, Keller AA, Weber K, Lorkowski S, et al. Complexity of fatty acid distribution inside human macrophages on single cell level using Raman micro-spectroscopy. Anal Bioanal Chem. 2014;406:7037–46.
[49] Matthaus C, Krafft C, Dietzek B, Brehm BR, Lorkowski S, Popp J. Noninvasive imaging of intracellular lipid metabolism in macrophages by Raman microscopy in combination with stable isotopic labeling. Anal Chem. 2012;84:8549–56.
[50] Naemat A, Elsheikha HM, Boitor RA, Notingher I. Tracing amino acid exchange during host-pathogen interaction by combined stable-isotope time-resolved Raman spectral imaging. Sci Rep. 2016;6:20811.

[51] El-Mashtoly SF, Yosef HK, Petersen D, Mavarani L, Maghnouj A, Hahn S, et al. Label-free Raman spectroscopic imaging monitors the integral physiologically relevant drug responses in cancer cells. Anal Chem. 2015;87:7297–304.

[52] Braeutigam K, Bocklitz T, Schmitt M, Roesch P, Popp J. Raman spectroscopic imaging for the real-time detection of chemical changes associated with docetaxel exposure. Chem Phys Chem. 2013;14:550–3.

[53] Salehi H, Derely L, Vegh AG, Durand JC, Gergely C, Larroque C, et al. Label-free detection of anticancer drug paclitaxel in living cells by confocal Raman microscopy. Appl Phys Lett. 2013;102:113701.

[54] Salehi H, Middendorp E, Panayotov I, Dutilleul PY, Vegh AG, Ramakrishnan SK, et al. Confocal Raman data analysis enables identifying apoptosis of MCF-7 cells caused by anticancer drug paclitaxel. J Biomed Opt. 2013;18:056010.

[55] Schie IW, Alber L, Gryshuk AL, Chan JW. Investigating drug induced changes in single, living lymphocytes based on Raman micro-spectroscopy. Analyst. 2014;139:2726–33.

[56] Draux F, Gobinet C, Sule-Suso J, Manfait M, Jeannesson P, Sockalingum GD. Raman imaging of single living cells: probing effects of non-cytotoxic doses of an anti-cancer drug. Analyst. 2011;136:2718–25.

[57] Huang H, Shi H, Feng SY, Chen W, Yu Y, Lin D, et al. Confocal Raman spectroscopic analysis of the cytotoxic response to cisplatin in nasopharyngeal carcinoma cells. Anal Meth. 2013;5:260–6.

[58] El-Mashtoly SF, Petersen D, Yosef HK, Mosig A, Reinacher-Schick A, Kötting C, et al. Label-free imaging of drug distribution and metabolism in colon cancer cells by Raman microscopy. Analyst. 2014;139:1155–61.

[59] Cheng JX, Xie XS. Coherent Raman scattering. Boca Raton, FL, USA: CRC Press, 2012.

[60] Wei D, Chen S, Liu Q. Review of fluorescence suppression techniques in Raman spectroscopy. Appl Spectrosc Rev. 2015;50:387–406.

[61] Diem M, Mazur A, Lenau K, Schubert J, Bird B, Miljkovic M, et al. Molecular pathology via IR and Raman spectral imaging. J Biophotonics. 2013;6:855–86.

[62] Krafft C, Codrich D, Pelizzo G, Sergo V. Raman and FTIR microscopic imaging of colon tissue: a comparative study. J Biophoton. 2008;1:154–69.

[63] Krafft C, Belay B, Bergner N, Bergner N, Romeike BF, Reichart R, et al. Advances in optical biopsy–correlation of malignancy and cell density of primary brain tumors using Raman microspectroscopic imaging. Analyst. 2012;137:5533–7.

[64] Bergner N, Krafft C, Geiger KD, Kirsch M, Schackert G, Popp J. Unsupervised unmixing of Raman microspectroscopic images for morphological analysis of non-dried brain tumor specimens. Anal Bioanal Chem. 2012;403:719–25.

[65] Bergner N, Medyukhina A, Geiger KD, Kirsch M, Schackert G, Krafft C, et al. Hyperspectral unmixing of Raman micro-images for assessment of morphological and chemical parameters in non-dried brain tumor specimens. Anal Bioanal Chem. 2013;405:8719–28.

[66] Bergner N, Bocklitz T, Romeike BF, Reichart R, Kalff R, Krafft C, et al. Identification of primary tumors of brain metastases by Raman imaging and support vector machines. Chemom Intell Lab Syst. 2012;117:224–32.

[67] Gajjar K, Heppenstall LD, Pang W, Ashton KM, Trevisan J, Patel II, et al. Diagnostic segregation of human brain tumours using Fourier-transform infrared and/or Raman spectroscopy coupled with discriminant analysis. Anal Meth. 2013;5:89–102.

[68] Abramczyk H, Brozek-Pluska B, Surmacki J, Jablonska-Gajewicz J, Kordek R. Raman 'optical biopsy' of human breast cancer. Prog Biophys Mol Biol. 2012;108:74–81.

[69] Mavarani L, Petersen D, El-Mashtoly SF, Mosig A, Tannapfel A, Kötting C, et al. Spectral histopathology of colon cancer tissue sections by Raman imaging with 532 nm excitation

provides label free annotation of lymphocytes, erythrocytes and proliferating nuclei of cancer cells. Analyst. 2013;138:4035–9.
[70] Horsnell JD, Smith JA, Sattlecker M, Sammon A, Christie-Brown J, Kendall C, Stone N. Raman spectroscopy - A potential new method for the intra-operative assessment of axillary lymph nodes. Surgeon-J Royal Colleges Surgeons Edinburgh Ireland. 2012;10:123–7.
[71] Bielecki C, Bocklitz TW, Schmitt M, Krafft C, Marquardt C, Gharbi A, et al. Classification of inflammatory bowel diseases by means of Raman spectroscopic imaging of epithelium cells. J Biomed Opt. 2012;17:076030.
[72] Minamikawa T, Harada Y, Koizumi N, Okihara K, Kamoi K, Yanagisawa A, et al. Label-free detection of peripheral nerve tissues against adjacent tissues by spontaneous Raman microspectroscopy. Histochem Cell Biol. 2013;139:181–93.
[73] Cals FL, Schut TC, Hardillo JA, de Jong RJ, Koljenovic S, Puppels GJ. Investigation of the potential of Raman spectroscopy for oral cancer detection in surgical margins. Lab Invest. 2015;95:1186–96.
[74] Stewart S, Kirschner H, Treado PJ, Priore R, Tretiakova M, Cohen JK. Distinguishing between renal oncocytoma and chromophobe renal cell carcinoma using Raman molecular imaging. J Raman Spectrosc. 2014;45:274–80.
[75] Marzec KM, Wrobel TP, Rygula A, Maslak E, Jasztal A, Fedorowicz A, et al. Visualization of the biochemical markers of atherosclerotic plaque with the use of Raman, IR and AFM. J Biophotonics. 2014;7:744–56.
[76] Lattermann A, Matthäus C, Bergner N, Beleites C, Romeike BF, Krafft C, et al. Characterization of atherosclerotic plaque depositions by Raman and FTIR imaging. J Biophotonics. 2013;6:110–21.
[77] Pilarczyk M, Mateuszuk L, Rygula A, Kepczynski M, Chlopicki S, Baranska M, et al. Endothelium in spots - high-content imaging of lipid rafts clusters in db/db mice. PLoS One. 2014;9:e106065.
[78] Kong K, Rowlands CJ, Varma S, Perkins W, Leach IH, Koloydenko AA, et al. Diagnosis of tumors during tissue-conserving surgery with integrated autofluorescence and Raman scattering microscopy. Proc Natl Acad Sci USA. 2013;110:15189–94.
[79] Kong K, Zaabar F, Rakha E, Ellis I, Koloydenko A, Notingher I. Towards intra-operative diagnosis of tumours during breast conserving surgery by selective-sampling Raman micro-spectroscopy. Phys Med Biol. 2014;59:6141–52.
[80] Patil CA, Kirshnamoorthi H, Ellis DL, van Leeuwen TG, Mahadevan-Jansen A. A clinical instrument for combined Raman spectroscopy-optical coherence tomography of skin cancers. Lasers Surg Med. 2011;43:143–51.
[81] Ashok PC, Praveen BB, Bellini N, Riches A, Dholakia K, Herrington CS. Multi-modal approach using Raman spectroscopy and optical coherence tomography for the discrimination of colonic adenocarcinoma from normal colon. Biomed Opt Express. 2013;4:2179–86.
[82] Bocklitz T, Braeutigam K, Urbanek A, Hoffman F, von Eggeling F, Ernst G, et al. Novel workflow for combining Raman spectroscopy and MALDI-MSI for tissue based studies. Anal Bioanal Chem. 2015;407:7865–73.

Anastasia Rousaki, Luc Moens and Peter Vandenabeele

8 Archaeological investigations (archaeometry)

Abstract: Archaeometry is the research area on the edge between humanities and natural sciences: it uses and optimises methods from chemistry, spectroscopy, physics, biology, etc. to help answering research questions from humanities. In general, these objects are investigated for several reasons. Besides the fundamental interest to know about the materials that were used in the past, the study of artefacts can support their preservation, either by helping to select optimal storage or display conditions, either by investigating decay pathways and suggesting solutions. Other reasons for art analysis include provenance studies, dating the artefact or identifying forgeries. Since several years, Raman spectroscopy is increasingly applied for the investigation of objects of art or archaeology. The technique is well-appreciated for the limited (or even absent) sample preparation, the relative straightforward interpretation of the spectra (by fingerprinting – comparing them against a database of reference pigments) and its speed of analysis. Moreover, the small spectral footprint – allowing to record a molecular spectrum of particles down to 1 µm, the typical size of pigment grains – is certainly a positive property of the technique. Raman spectroscopy can be considered as rather versatile, as inorganic as well as organic materials can be studied, and as the technique can gather information on crystalline as well as on non-crystalline phases. As a consequence, Raman spectroscopy can be used to study antique objects and twentieth-century synthetic (organic) materials – illustrating the wide range of applications. Finally, the technique is as non-destructive, provided the laser power is kept sufficiently low not to damage the artwork. In literature, the terms “non-invasive” and “non-destructive” are used, where the first term means that no sampling is involved, and the latter term indicates that no sample is taken or that during analysis the sample is not consumed (destroyed) and remains available for further analysis.

Keywords: archaeometry, Raman spectroscopy, art analysis, cultural heritage research

8.1 Introduction

Archaeometry is the research area on the edge between humanities and natural sciences: it uses and optimises methods from chemistry, spectroscopy, physics, biology,

This article has previously been published in the journal *Physical Sciences Reviews*. Please cite as: Rousaki, A., Moens, L., Vandenabeele, P. Archaeological investigations (archaeometry) *Physical Sciences Reviews* [Online] **2019**, 4. DOI: 10.1515/psr-2017-0048.

https://doi.org/10.1515/9783110515312-008

etc. to help answering research questions from humanities. In general, these objects are investigated for several reasons. Besides the fundamental interest to know about the materials that were used in the past, the study of artefacts can support their preservation, either by helping to select optimal storage or display conditions, either by investigating decay pathways and suggesting solutions. Other reasons for art analysis include provenance studies, dating the artefact or identifying forgeries.

Since several years, Raman spectroscopy is increasingly applied for the investigation of objects of art or archaeology [1–3]. The technique is well-appreciated for the limited (or even absent) sample preparation, the relative straightforward interpretation of the spectra (by fingerprinting – comparing them against a database of reference pigments) and its speed of analysis. Moreover, the small spectral footprint – allowing to record a molecular spectrum of particles down to 1 µm, the typical size of pigment grains – is certainly a positive property of the technique. Raman spectroscopy can be considered as rather versatile, as inorganic as well as organic materials can be studied, and as the technique can gather information on crystalline as well as on non-crystalline phases. As a consequence, Raman spectroscopy can be used to study antique objects and twentieth-century synthetic (organic) materials – illustrating the wide range of applications [1]. Finally, the technique is as non-destructive, provided the laser power is kept sufficiently low not to damage the artwork. In literature, the terms "non-invasive" and "non-destructive" are used, where the first term means that no sampling is involved, and the latter term indicates that no sample is taken or that during analysis the sample is not consumed (destroyed) and remains available for further analysis.

In the next paragraphs, we will discuss some recent applications of Raman spectroscopy in archaeometry. First, an overview will be given of some novel trends involving *in vitro* (laboratory) research of samples or small objects. Secondly, non-invasive applications, involving mobile Raman instrumentation, will be discussed, followed by a discussion on how to combine Raman spectroscopy with other techniques.

8.2 Benchtop Raman instrumentation and possibilities

New applications in Raman spectroscopy were often driven by instrumental changes. Thus, the introduction of confocal Raman microscopy and the application of Charge-Coupled Device (CCD) detectors have triggered the application of Raman spectroscopy in art and archaeology [4]. In general, benchtop Raman instruments share some favourable characteristics such as the ability of coupling multiple monochromatic light sources (single, dual or multiple laser systems), optimal wavenumber stability, high spectral resolution (depending on the laser excitation within the same configuration), focusing through high magnification objectives and incorporation of calibrated motorised stages.

Amongst the first applications of Raman spectroscopy, pigment identification in mediaeval manuscripts [5–7] was an important application, showing the different possibilities of the technique. Soon, people realised that this approach could be extended towards other applications, including other paintings [8], corrosion products [9], gems [10], glass [11] and ceramics [12]. The artefacts, that can be studied via Raman spectroscopy date from prehistory till contemporary art. A credible example is the chemical characterization of rock art paintings, which are considered as one the most ancient forms of human expression. In general, rock art is considered as the painted or engraved or carved natural rock surfaces made by indigenous people. The pictographs or petroglyphs found on stone surfaces can stylistically vary from abstract to geometric or more naturalistic and can be can be found in Europe, Africa, America, Latin-America, Asia, and Oceania. Their preservation and documentation is considered highly important, in order to understand the technological level of the societies that created them. Several studies can be found in the literature highlighting the use of Raman spectroscopy on the characterization of the mainly red, yellow green, white and black pigmented layers [13, 14]. Considering their preservation state, Raman spectroscopy is employed to study (bio)degradation products associated with weathering and human activity. For modern art studies, Raman spectroscopy was used for the investigation of azo pigments [15] and copper phthalocyanine (PB15). The latter is often identified as an important pigment in twentieth-century artworks. However, PB15 is used in different polymorphic forms (α-CuPc, β-CuPc and ε-CuPc). Raman spectroscopy is a very valuable technique for the detection of this pigment in paint systems and it is able to distinguish polymorphs [16]. The discrimination of PB15 polymorphs can be used to date and authenticate artworks [16] as they appeared at different times on the paint market [17, 18]. The Raman spectra of α-CuPc and β-CuPc differ only in relative intensities of some bands. Furthermore, using linear discriminant analysis (LDA), the α-, β- and ε-CuPc pigments could be discriminated using the Raman intensity ratios as variables [16].

The approach was also increasingly applied for the study of natural organic products, although, due to fluorescence suppression, this was often related to Fourier-transform (FT-) applications. One can think that by shifting to near-IR excitation lasers (typically 1064 nm), the contribution of fluorescence is partially avoided. However, simultaneously the scattering capability is as well decreasing, and as a consequence, higher radiance together with larger measuring times should be applied, with the risk of overheating the sample [19, 20]. Moreover, the use of FT-Raman technology implies the use of solid state detectors (usually low-bandgap semiconductors), as silica based CCDs detectors are less sensitive in the near IR region (for wavelengths larger than 1.1 µm) [20]. Despite possible drawbacks, FT-Raman spectroscopy is a very powerful tool with many application in cultural heritage, such as (among others) the study of native American Indian rock art [21], characterization of Egyptian pigments [22] investigation of thioindigo–palygorskite mixtures resembling Maya blue [23] etc. Furthermore, elaborated FT-Raman spectral databases can be found in the literature [24, 25].

Today, Raman spectroscopy has proven to be versatile approach in archaeometry, and it is often considered as a first choice technique when it comes to art analysis. Unfortunately, compared to other spectroscopic methods, the Raman effect is relatively weak. On the other hand, it has an excellent spatial resolution, and by focussing the laser beam on, for instance, a pigment grain, it is possible to obtain good quality spectra. Unfortunately, not all molecules yield intense Raman spectra. However, when the laser wavelength can be selected, one can choose a laser that has a frequency close to the electronic absorption region, thus enhancing the signal with several orders of magnitude. This resonance enhancement produced, for instance, interesting results when studying lazurite, ultramarine blue, chrome based pigments [26, 27] and carotenoids (including corals) [28–30]. Although, lasers that are situated at the blue and green region of the electromagnetic spectrum are considered optimal for Raman signal enhancement of carotenoids [31], red and near IR lasers can also be used for the identification of these molecules [revealing at least the two strongest Raman bands (in-phase $\nu(C=C)$ and $\nu(C{-}C)$ stretching vibrations) and one medium intensity feature (in-plane $\rho(C{-}CH_3)$ rocking mode) [31]] even sometimes with portable Raman instrumentation [32]. However, when selecting a laser, it is important to consider that the intrinsic Raman intensity is proportional to the fourth power of the laser frequency. For resonance Raman spectroscopy, the laser wavelength should be in agreement with the vibrations of the target molecule under study. On the other hand, the spectroscopist should be aware that theoretically when selecting shorter, higher energetic wavelengths, fluorescence is increasing even though the Raman signal is enhancing.

Moreover, some materials, such as dyes, are not present in high local concentrations, but are rather present in a dispersed way, which makes it difficult to focus the laser on the sample. Surface enhanced Raman spectroscopy (SERS) [33, 34] can be of help for identifying the dye molecules on hand. Indeed, by adsorbing analyte molecules to noble metal nanostructures, the observed Raman signal can be seriously enhanced, while the fluorescence is reduced. This is achieved by combined electromagnetic [35] and charge-transfer [36] mechanisms. Silver nanoparticles can be produced in different ways. In cultural heritage research, often silver nanoparticles are obtained by reduction of silver nitrate by using citrate [37]. Another approach starts from silver sulphate that is reduced by glucose, under microwave irradiation [38]. In that case, very reproducible results could be obtained, while sodium citrate is used as capping agent. Also photoreduction [39] and laser-ablation [40] were proposed as promising alternatives for the production of silver nanoparticles. Silver island films (AgIFs) or thin films of silver over nanospheres were also proposed as SERS substrates. In that case, these substrates were either directly deposited [41] on the coloured surface, or firstly adsorbed to a silver-alumina support [42], on which the analyte afterwards was adsorbed.

The technique was discovered in 1974 [33], but its application in artworks started only in 1987 [43] by the identification of madder in historical textile [44]. SERS proved to be very successful for the identification of pure, commercial dyes, detection of natural organic colorants [38, 44] such as anthraquinones, flavonoids, naftoquinones, and rhodamines, as they are the main components in dyes found on works of art. However, some challenges should be taken into account. For lake pigments or mordant dyes, hydrolysis treatment is the first step towards the extraction of the dye from the inorganic substrate or the textile fibre, respectively [20].

Raman spectroscopy is known to be a surface sensitive technique as the molecular information that is retrieved during Raman experiments typically originates from the top layer, since the (visual) laser light does not penetrate deep into the sample. When using a confocal set-up it is possible to focus through a transparent layer and thus to obtain information from deeper areas. If the surface layer is not transparent one typically has to rely on embedding a sample and cutting and polishing it, to achieve an overview of the different layers. However, as in many cases sampling is considered undesirable (damage, time consuming sample preparation, contamination, etc.), in 2005, spatially offset Raman spectroscopy (SORS) was proposed by Matousek et al. [45, 46]. By collecting the Raman signal from an area that is spatially offset from the laser spot, it is possible to retrieve molecular information from the subsurface underneath a highly turbid top layer [47]. The larger a spatial offset is selected, the deeper the information is retrieved, but consequently the weaker the collected signal is. Typically high power infrared lasers are used with a spatial offset of few millimetres and penetration depths up to several cm. SORS has, amongst others, proven to be successful for analysing pharmaceuticals through the packaging [48], cancer research [49] and other medical applications [50].

In works of art, the painted layers are typically few micrometres thick. In order to achieve non-invasive stratigraphic analysis, an alternative to the SORS technique was proposed namely, microscale-spatially offset Raman spectroscopy (micro-SORS). This approach is favourable for research of cultural heritage objects [51]. In micro-SORS, laser power, spot size and collection area are reduced, as well as the (small) spatial offset, which allows to combine the advantageous lateral resolution of Raman microscopy with the retrieval of molecular information on the stratigraphy. Although the transition from SORS to micro-SORS might seem a simple case of miniaturisation, spatial constraints apply. Moreover, the experimental set-up for *full* micro-SORS requires the laser to be focused on the sample through a different lens as the collection of the Raman signal, which is not directly compatible with commercial Raman instrumentation [51]. Therefore, alternative approaches were proposed, such as *defocusing* micro-SORS, which uses a commercially available Raman microscope. During this approach, the spatial offset is obtained by defocusing the (collection) optics, and thus probing an area which contains a large fraction of the spatially offset signal [52, 53]. The more defocused the spectra are, the more information is obtained from deeper layers; spectral post-processing involves scaled subtraction of the in-focus spectrum from the out-of-focus one to diminish the contribution of the surface [20]. It should be remarked that, despite being straightforward in use, *defocusing* micro-SORS does not yield optimal separation between layers, as can be achieved with *full* micro-SORS. To overcome some of the practical constraints of *full* micro-SORS, *fibre-optics* micro-SORS was recently proposed [54]. The technique utilizes two parallel glass fibres that are in contact with the sample. The excitation fibre is mounted on a non-moving part, whereas the collection fibre is mounted on a translation stage, allowing to set the spatial offset. *Fibre-optics* micro-SORS was successfully demonstrated on compounds with thicknesses of the top layer ranging from 50 to 500 μm [54]. Despite this successful proof-of-concept, this approach should still be further optimised (e. g. tackling fibre brittleness, optimal fibre thickness, applying filters).

The applicability of the micro-SORS technique, in respect to cultural heritage studies, is demonstrated on thin layered mock up samples [53] and real artefacts such as, stucco and terracotta sculptures and painted plaster in combination with cross-section analysis [55]. Moreover, micro-SORS chemical interrogations on mock up (among others) samples consist of examining the sequence of which the layers are made [56] and if two components are situated within the same or different layers [57].

Laboratory Raman spectrometers are coupled with calibrated automated positioning stages, allowing the movement on the XY or XYZ axis. This characteristic makes the micro-Raman spectrometers suitable for retrieving molecular mappings and thus connecting chemical information with its spatial distribution. By focusing the objective in one point on the sample, the automated stage can be moved inside a preset, well-defined area collecting sequences of Raman spectra, stored as sequential point measurements (movement in XY axis). An advantage of acquiring full spectra is that the construction of the Raman mapping image can be decided upon the need of analysis. For surfaces that are not completely flat, an autofocussing option (movement in Z axis) can be applied for each analysed point. Autofocusing is a very crucial characteristic when mapping cultural heritage objects and/or cross sections that are not completely flat. It increases the obtained spectral quality and thus the obtained image. Lofrumento et al. [58] used Raman mapping in a study dedicated to Ethiopian prehistoric rock painting, for the investigation of the possible origin of calcium oxalate. Conti et al. [59] applied micro-Raman mapping for the study of calcium oxalate films from samples taken from the north façade of the Cathedral of Trento. For subsurface Raman mapping, recently, two studies explored the possibilities of *full* micro-SORS mapping on samples mimicking real situations found in artworks with hidden layers [60] and *defocusing* micro-SORS mapping on a nineteenth-century porcelain card [61].

Point-to-point mapping can consume several working hours (even more increased when autofocussing is involved). Linear lasers can drastically decrease the overall measuring time by simultaneously recording series of points in a row.

Opposite to mapping, Raman imaging uses larger illumination areas and filters to obtain the desired image usually acquired by the selection of one component, e. g. a specific Raman band [62]. Raman imaging seems abandoned in cultural heritage studies, as Raman mapping can produce more detailed information of the area under investigation or even the entire dataset can be used for chemometrical analysis.

8.3 Direct Raman analysis

Raman spectroscopy is considered traditionally as a non-destructive technique as the laser power can be kept low in order to avoid alteration and permanent thermal damage on the object/sample. Technological evolutions and development of fibre optics brought Raman spectrometers out of the laboratory, into the field or inside museums. The latter helped towards the improvement of direct, non-invasive analysis as large scale monuments or art works started to be investigated in their natural environment or in dedicated museum collections without sampling and thus jeopardising the artefact. Compared with benchtop Raman spectrometers, instruments dedicated to the *in situ* analysis have in general, a decreased wavenumber stability (due to vibrations, thermal fluctuations etc.), thus they need frequent calibration. They usually have a larger spectral footprint and worse spectral resolution and focusing is achieved through the evaluation of signal-to-noise ratio, the number of Raman bands, their shapes and intensities in the spectra of the unknown in a given position or through fixed focal distances (when using light blockers).

Before discussing the use of mobile Raman spectrometers in archaeometrical research, one should be aware of definitions found in literature about not only the kind of analysis performed but also the kind of the instrument used. On site or in the field or *in situ* analysis is proposed when the study is performed in the environment where the work of art is found or displayed, whereas non-invasive is referring to the fact that no sampling is necessary [20, 63–65]. A wide range of definitions exist for characterizing the mobility of a Raman instrument. Previous works [63–65] distinguished the instruments in: (i) Transportable: instruments, usually not designed for mobility. Benchtop Raman instruments can also be referred as transportable; (ii) Mobile: a general term for stable spectrometers designed for mobility; (iii) Portable: compact, usually battery-operated Raman spectrometers that can be carried by a single person; (iv) Handheld: low weight spectrometers, held while performing measurement on one hand of a person; (v) Palm: ultra-light, ultra-mobile Raman spectrometers.

The first attempts in analysing works of art with fibre optics probe heads was done in the case of easel paintings by probing through the top varnish layers [66]. Fibre optics where attached to a FT-Raman instrument allowing direct analysis on the artefact. Following this, in 2004 [67], a new more compact portable Raman instrument was introduced, coupled with a 785 nm diode laser, camera and adjustable power. This instrument allowed for the direct, *in situ* Raman analysis as demonstrated successfully by investigating the tomb of Menna, Theban Necropolis, Egypt [68], folios from the breviary of Arnold of Egmond [69], vault paintings in the Antwerp cathedral [70] etc. In 2007 [71], a short comparative review on commercially available Raman instruments was published.

In cultural heritage studies, portable Raman spectrometers are more favourable due to their coupling with long fibre optics cables which allows the remote investigation of art objects or the easy access to larger/bigger artefacts. Moreover, such Raman instruments allow the freedom to the operator to choose the spectral and measuring parameters. For the handheld and ultra-mobile instruments the probe head is directly attached to the small spectrometer. These types of instruments are usually discussed in mineralogy [72] but also in the gems and gemstones [73] literature.

Portable Raman spectrometers are equipped with one or two lasers, while handheld instruments have one laser and a fixed spectral range. In 2015, Bruker introduced a new handheld instrument on the market, using different lasers and a shifted wavelength algorithm to correct for fluorescence background, with low laser power and a wide spectral range [74, 75]. The instrument was evaluated on laboratory specimens and painted sculptures [74], on pigments, organic materials and binders in modern art [76] but also on mineral and carbons relevant to exobiology and geobiology [75].

Mobile Raman spectroscopy has grown to be a first choice technique in many cultural heritage studies. Besides wall painting analysis [63], mobile Raman spectroscopy found a remarkable application in the study of prehistoric rock art paintings. Raman researchers investigated many rock sites situated around the world, e. g. South Africa [77], France [78], India [79], Spain [80–82], Patagonia, Argentina [32]. For the former two studies, a rather transportable instrument was used [32]. Other applications include the direct analysis of glyptics [83], paintings and mortars from the Marcus Lucretius Pompeian house [84], stone building materials under environmental influence [85], one site analysis of enamel porcelains and stoneware [86, 87] and so on.

Opposite to benchtop instruments, focusing the probe heads of portable spectrometers is rather challenging. Without visual inspection through coupled objective lenses, one should rely on the evaluation of the Raman spectra. Positioning systems need to assure stability and thus optimal focusing. They are usually tripods, articulating arms etc. For works of art that are situated on higher positions above the ground, the probe heads can be positioned on scaffoldings or long fibre optics cables can be mounted on extended tripods. If, bringing positioning accessories in the field is rather time consuming and sometimes limits the mobility, cups with fixed focal distances can be used [32, 63]. These cups (tip covered with protective foams) are slided over the probe heads and measuring of the surface is achieved by placing them in contact with the surface under investigation. Moreover, the cups serve also as light blockers (blocking the signal of ambient light) when working in open air environments during daylight.

As the Raman signal, suffers environmental interference it is advised to perform *in situ* measurements in complete darkness or during night. When studying various artistic objects with multiple techniques, e. g. in a museum, or when the measurements are situated during restoration campaigns or when the analysis is carried out on public hours, double cloth tents can be used and the object together with the Raman spectrometer can be placed inside them.

Although benchtop Raman spectrometers are often performing better compared to their mobile counterparts, mobile/portable/handheld Raman instruments are used extensively on the field of cultural heritage, because of their non-invasive characteristics. Recently, an *in situ* Raman mapping configuration is published in the literature, for the analysis of porcelain cards [88]. Even though the results were very promising several milestones still should be reached in order for the technique to be widely applicable. Furthermore, the firsts steps, for subsurface mobile Raman mapping, are taken by the development of *defocusing* micro-SORS in a mobile set-up [89] or the development of a prototype of a *full* micro-SORS probe head [90], both suggesting the possibilities for *in situ* experiments in several fields of application, including archaeometry.

8.4 Raman spectroscopy in combination with other techniques

Elaborated physicochemical protocols for cultural heritage analysis often use a wide range of techniques. Artistic artefacts are complex compositions that require both elemental, molecular and separation techniques in order for the different components to be identified. The analytical approaches can yield qualitative or quantitative results. Ideally, the documentation should be as less invasive as possible. The evaluation of the artefacts can carried out with several macro-imaging techniques, such as X-ray radiography, infrared reflectography, 2D and 3D digital microscopy etc. If the ethics and the state of the object allow the sampling then invasive, laboratory techniques can be employed. For laboratory measurements, the analysis should be non-destructive and thus, the same sample needs to be investigated with a variation of techniques and finally kept for future studies. If destructive techniques (with sample consumption) are employed these should be scheduled towards the end of the analytical protocol.

Since combined methodological approaches are usually preferable, Raman spectroscopy can be used in combination with other molecular or elemental analytical techniques. One popular approach is the combination with Fourier-transform infrared spectroscopy (FTIR) [91]. This technique gives rise to vibrations due to a change in polarization, while Raman spectroscopy requires changes in polarisability [62]. In general, intense infrared active modes are weak in Raman spectroscopy except if, in some cases, structural changes occur inside the matrices e. g. lowering of crystallinity, distortion of structures etc. FTIR spectroscopy is suitable for both organic binders/varnishes and inorganic pigments [92]. Moreover, it can be applied non-invasively with reflectance infrared portable instruments [93]. FTIR analysis is sometimes combined with thermal analysis methods for discrimination of historic plasters and mortars [94].

X-ray diffraction (XRD) is another technique that is highly suitable for the analysis of crystalline phases and that can complement Raman spectroscopy for the analysis of inorganic molecules. For example, the identification of feldspars can be sometimes challenging with Raman spectroscopy, as high spectral resolution and exact wavenumber calibration are necessary to identify the correct end member. Portable XRD (p-XRD) instruments can also be used for the analysis of cultural heritage objects [95] but in some cases positioning and alignment can be quite time consuming procedures [96].

Scanning electron microscopy coupled with energy dispersive X-ray spectroscopy (SEM-EDS) allows for the characterization of the morphology and the elemental distribution of cross sections or untreated samples, or even small objects. Depending on the electrical conductivity of the sample, it should be coated or measured as it is. From the coupling between SEM and a Raman spectrometer a hybrid system immerged, namely the structural and chemical analyser (SCA) [97]. In cultural heritage research such system was used for the identification of degradation products of archaeological guilded irons [97].

One of the most discussed combinations is the coupling of X-ray fluorescence research with Raman spectroscopy in cultural heritage studies. Thus molecular and elemental information is combined. Although the complementation is considered successful, one should be aware of the different sampling volumes and depths of both techniques when interpreting the results. Visual light is not penetrating deeper than the surface while X-rays can sometimes penetrate through the sample and/object. For the latter, the information received is a sum of the same or different elements found in the entire stratigraphy. In archaeometry, micro-Raman and micro-XRF can be combined [98] but also their mobile counterparts [99]. Analysis of powdered samples or scrapings of art allow the use of another XRF technique [100], namely total reflection X-ray fluorescence (TXRF) for fast and sensitive (down to trace elements) analysis. Combined XRF-XRD [101] and XRF-Raman [102] mobile configurations are also reported in the archaeometrical research field. The evolution of macro-XRF (MA-XRF) imaging and its applications on art analysis are well established and demonstrated in numerous published studies [103].

8.5 Conclusions

Raman spectroscopy has grown to be one of the most appreciated techniques in the study of art and archaeology. Technical advantages shaped the technique and moreover, initiated the expected development of more novel Raman techniques the following years. For cultural heritage studies, advanced approaches involve the development of new methods such as micro-SORS, new instrumentation such as dispersive instruments with IR laser, enhancement of Raman signals for optimal identification, upgraded benchtop and portable systems, subsurface and *in situ* molecular mappings, and so on. Raman spectroscopy seems to be the perfect technique for novel developments in the interesting research field of archaeometry.

References

[1] Vandenabeele P, Edwards HGM, Moens L. A decade of Raman spectroscopy in art and archaeology. *Chem Rev.* 2007;107:675–86.

[2] Edwards HGM, Chalmers JM, editors. Raman spectroscopy in archaeology and art history. UK, The Royal Society of Chemistry: Cambridge, 2005.

[3] Colomban P. The on-site/remote Raman analysis with mobile instruments: A review of drawbacks and success in cultural heritage studies and other associated fields. J Raman Spectrosc. 2012;43:1529–35.

[4] Dhamelincourt P, Wallart F, Leclercq M, Nguyen AT, Landon DO. Laser Raman molecular microprobe (MOLE). Anal Chem. 1979;51:414A21A.

[5] Guineau B, Coupry C, Gousset MT, Forgerit JP, Vezin J. Identification de bleu de lapis-lazuli dans six manuscrits à peintures du XIIe siècle provenant de l'abbaye de Corbie. Scriptorium. 1986;40:157–71.

[6] Clark RJH. Raman microscopy: Application to the identification of pigments on medieval manuscripts. Chem Soc ReV. 1995;24:187–96.

[7] Vandenabeele P, Wehling B, Moens L, Dekeyzer B, Cardon B, Von Bohlen A, et al. Pigment investigation of a late-medieval manuscript with total reflection X-ray fluorescence and micro-Raman spectroscopy. Analyst. 1999;124:169–72.

[8] Andalo C, Bicchieri M, Bocchini P, Casu G, Galletti GC, Mando PA, et al. The beautiful "Trionfo d'Amore" attributed to Botticelli: A chemical characterisation by proton-induced X-ray emission and micro-Raman spectroscopy. Anal Chim Acta. 2001;429:279–86.

[9] Hayez V, Guillaume J, Hubin A, Terryn H. Micro-Raman spectroscopy for the study of corrosion products on copper alloys: Setting up of a reference database and studying works of art. J Raman Spectrosc. 2004;35:732–38.

[10] Kiefert L, Hänni HA, Chalain JP, Weber W. Identification of filler substances in emeralds by infrared and Raman spectroscopy. J Gemmol. 1999;26:501–20.

[11] Colomban P, Schreiber HD. Raman signature modification induced by copper nanoparticles in silicate glass. J Raman Spectrosc. 2005;36:884–90.

[12] Clark RJH, Curri ML. The identification by Raman microscopy and X-ray diffraction of iron-oxide pigments and of the red pigments found on Italian pottery fragments. J Mol Struct. 1998;440:105–11.

[13] Rousaki A, Bellelli C, Carballido Calatayud M, Aldazábal V, Custo G, Moens L, et al. Micro-Raman analysis of pigments from Hunter–Gatherer archaeological sites of North Patagonia (Argentina). J Raman Spectrosc. 2015;46:1016–24.

[14] Hernanz A, Gavira-Vallejo JM, Ruiz-López JF, Edwards HGM. A comprehensive micro-Raman spectroscopic study of prehistoric rock paintings from the Sierra de las Cuerdas, Cuenca, Spain. J Raman Spectrosc. 2008;39:972–84.

[15] Vandenabeele P, Moens L, Edwards HGM, Dams R. Raman spectroscopic database of azo pigments and application to modern art studies. J Raman Spectrosc. 2000;31:509–17.

[16] Defeyt C, Van Pevenage J, Moens L, Strivay D, Vandenabeele P. Micro-Raman spectroscopy and chemometrical analysis for the distinction of copper phthalocyanine polymorphs in paint layers. Spectrochim Acta A. 2013;115:636–40.

[17] Moser FH, Thomas AL. Phthalocyanine compounds. USA, Reinhold Publishing Corporation: New York, 1963.

[18] Herbst W, Hunger K, Wilker G, Ohleier H, Winter R. Industrial organic pigments, 3rd ed. Weinheim FRG: Wiley-VCH Verlag GmbH & Co. KGaA, 2004

[19] Bersani D, Lottici PP. Raman spectroscopy of minerals and mineral pigments in archaeometry. J Raman Spectrosc. 2016;47:499–30.

[20] Bersani D, Conti C, Matousek P, Pozzi F, Vandenabeele P. Methodological evolutions of Raman spectroscopy in art and archaeology. Anal Methods. 2016;8:8395–99.

[21] Edwards HGM, Drummond L, Russ J. Fourier-transform Raman spectroscopic study of pigments in native American Indian rock art: Seminole Canyon. Spectrochim Acta Part A. 1998;54:1849–56.

[22] David R, Edwards HGM, Farwell DW, De Faria DL. Raman spectroscopic analysis of ancient Egyptian pigments. Archaeometry. 2001;43:461–73.

[23] Manciu FS, Ramirez A, Durrer W, Govani J, Chianelli RR. Spectroscopic analysis of a dye–Mineral composite–A Raman and FT-IR study. J Raman Spectrosc. 2008;39:1257–61.

[24] Castro K, Pérez-Alonso M, Rodríguez-Laso MD, Fernández LA, Madariaga JM. On-line FT-Raman and dispersive Raman spectra database of artists' materials (e-VISART database). Anal Bioanal Chem. 2005;382:248–58.

[25] Burgio L, Clark RJH. Library of FT-Raman spectra of pigments, minerals, pigment media and varnishes, and supplement to existing library of Raman spectra of pigments with visible excitation. Spectrochim Acta Part A. 2001;57:1491–21.

[26] Correia AM, Oliveira MJV, Clark RJH, Ribeiro MI, Duarte ML. Characterization of Pousão pigments and extenders by micro-X-ray diffractometry and infrared and Raman microspectroscopy. Anal Chem. 2008;80:1482–92.

[27] Colomban P. Lapis lazuli as unexpected blue pigment in Iranian Lâjvardina ceramics. J Raman Spectrosc. 2003;34:420–23.

[28] Ermakov IV, Sharifzadeh M, Ermakova M, Gellermann W. Resonance Raman detection of carotenoid antioxidants in living human tissue. J Biomed Opt. 2005;10:861–67.

[29] Bergamonti L, Bersani D, Csermely D, Lottici PP. The nature of the pigments in corals and pearls: A contribution from Raman spectroscopy. Spectrosc Lett. 2011;44:453–58.

[30] Gill D, Kilponen RG, Rimai L. Resonance Raman scattering of laser radiation by vibrational modes of carotenoid pigment molecules in intact plant tissues. Nature. 1970;227:743–44.

[31] Jehlička J, Edwards HGM, Oren A. Raman spectroscopy of microbial pigments. Appl Environ Microbiol. 2014;80:3286–95.

[32] Rousaki A, Vázquez C, Aldazábal V, Bellelli C, Carballido Calatayud M, Hajduk A, et al. The first use of portable Raman instrumentation for the *in situ* study of prehistoric rock paintings in Patagonian sites. J Raman Spectrosc. 2017;48:1459–67.

[33] Fleischmann M, Hendra PJ, McQuillan AJ. Raman spectra of pyridine adsorbed at a silver electrode. Chem Phys Lett. 1974;26:163–66.

[34] Aroca R. Surface-enhanced vibrational spectroscopy. UK, John Wiley & Sons: Chichester, 2006.

[35] Jeanmaire DL, Van Duyne RP. Surface Raman spectroelectrochemistry: Part I. Heterocyclic, aromatic, and aliphatic amines adsorbed on the anodized silver electrode. J Electroanal Chem. 1977;84:1–20.

[36] Albrecht MG, Creighton JA. Anomalously intense Raman spectra of pyridine at a silver electrode. J Am Chem Soc. 1977;99:5215–17.

[37] Lee PC, Meisel DJ. Adsorption and surface-enhanced Raman of dyes on silver and gold sols. J Phys Chem. 1982;86:3391–95.

[38] Leona M. Microanalysis of organic pigments and glazes in polychrome works of art by surface-enhanced resonance Raman scattering. Proc Natl Acad Sci USA. 2009;106:14757–62.

[39] Cañamares MV, Garcia-Ramos JV, Gómez-Varga JD, Domingo C, Sanchez-Cortes S. Ag nanoparticles prepared by laser photoreduction as substrates for in situ surface-enhanced Raman scattering analysis of dyes. Langmuir. 2007;23:5210–15.

[40] Cañamares MV, Garcia-Ramos JV, Sanchez-Cortes S, Castillejo M, Oujja M. Comparative SERS effectiveness of silver nanoparticles prepared by different methods: A study of the enhancement factor and the interfacial properties. J Colloid Interface Sci. 2008;326:103–09.

[41] Whitney AV, Van Duyne RP, Casadio F. An innovative surface-enhanced Raman spectroscopy (SERS) method for the identification of six historical red lakes and dyestuffs. J Raman Spectrosc. 2006;37:993–02.

[42] Chen K, Leona M, Vo-Dinh K-C, Yan F, Wabuyele MB, Vo-Dinh T. Application of surface-enhanced Raman scattering (SERS) for the identification of anthraquinone dyes used in works of art. J Raman Spectrosc. 2006;37:520–27.

[43] Guineau B, Guichard V Identification des colorants organiques naturels par microspectrometrie Raman de resonance et par effet Raman exalte de surface (SERS). In: ICOM Committee for Conservation: 8th triennial meeting, Sydney, Australia, 6–11 September, 1987 Preprints, The Getty Conservation Institute: Marina del Rey, CA, 1987, 2, 659–66.

[44] Casadio F, Leona M, Lombardi JR, Van Duyne R. Identification of organic colorants in fibers, paints, and glazes by surface enhanced Raman spectroscopy. Acc Chem Res. 2010;43:782–91.

[45] Matousek P, Clark IP, Draper ERC, Morris MD, Goodship AE, Everall N, et al. Subsurface probing in diffusely scattering media using spatially offset Raman spectroscopy. Appl Spectrosc. 2005;59:393–00.

[46] Matousek P, Morris MD, Everall N, Clark IP, Towrie M, Draper E, et al. Numerical simulations of subsurface probing in diffusely scattering media using spatially offset Raman spectroscopy. Appl Spectrosc. 2005;59:1485–92.

[47] Buckley K, Matousek P. Non-invasive analysis of turbid samples using deep Raman spectroscopy. Analyst. 2011;136:3039–50.

[48] Eliasson C, Matousek P. Noninvasive authentication of pharmaceutical products through packaging using spatially offset Raman spectroscopy. Anal Chem. 2007;79:1696–01.

[49] Stone N, Baker R, Rogers K, William Parker A, Matousek P. Subsurface probing of calcifications with spatially offset Raman spectroscopy (SORS): Future possibilities for the diagnosis of breast cancer. Analyst. 2007;132:899–05.

[50] Matousek P, Stone N. Development of deep subsurface Raman spectroscopy for medical diagnosis and disease monitoring. Chem Soc Rev. 2016;45:1794–02.

[51] Conti C, Realini M, Colombo C, Matousek P. Comparison of key modalities of micro-scale spatially offset Raman spectroscopy. Analyst. 2015;140:8127–33.

[52] Matousek P, Conti C, Realini M, Colombo C. Micro-scale spatially offset Raman spectroscopy for non-invasive subsurface analysis of turbid materials. Analyst. 2016;141:731–39.

[53] Conti C, Colombo C, Realini M, Zerbi G, Matousek P. Subsurface Raman analysis of thin painted layers. Appl Spectrosc. 2014;68:686–91.

[54] Vandenabeele P, Conti C, Rousaki A, Moens L, Realini M, Matousek P. Development of a fiber-optics microspatially offset Raman spectroscopy sensor for probing layered materials. Anal Chem. 2017;89:9218–23.

[55] Conti C, Colombo C, Realini M, Matousek P. Subsurface analysis of painted sculptures and plasters using micrometre-scale spatially offset Raman spectroscopy (micro-SORS). J Raman Spectrosc. 2015;46:476–82.
[56] Conti C, Realini M, Colombo C, Sowoidnich K, Afseth NK, Bertasa M, et al. Noninvasive analysis of thin turbid layers using microscale spatially offset Raman spectroscopy. Anal Chem. 2015;87:5810–5.
[57] Conti C, Realini M, Botteon A, Colombo C, Noll S, Elliott SR, et al. Analytical capability of defocused μ-SORS in the chemical interrogation of thin turbid painted layers. Appl Spectrosc. 2016;70:156–61.
[58] Lofrumento C, Ricci M, Bachechi L, De Feo D, Castellucci EM. The first spectroscopic analysis of Ethiopian prehistoric rock painting. J Raman Spectrosc. 2012;43:809–16.
[59] Conti C, Aliatis I, Colombo C, Greco M, Possenti E, Realini M, et al. μ-Raman mapping to study calcium oxalate historical films. J Raman Spectrosc. 2012;43:1604–11.
[60] Botteon A, Conti C, Realini M, Colombo C, Matousek P. Discovering hidden painted images: Subsurface imaging using microscale spatially offset Raman spectroscopy. Anal. Chem. 2017;89:792–98.
[61] Rousaki A, Botteon A, Colombo C, Conti C, Matousek P, Moens L, et al. Development of defocusing micro-SORS mapping: A study of a nineteenth century porcelain card. Anal Methods. 2017;9:6435–42.
[62] Vandenabeele P. Practical Raman spectroscopy: An introduction. UK, John Wiley & Sons: Chichester, 2013.
[63] Lauwers D, Garcia Hutado A, Tanevska V, Moens L, Bersani D, Vandenabeele P. Characterisation of a portable Raman spectrometer for in situ analysis of art objects. Spectrochim Acta Part A. 2014;118:294–01.
[64] Vandenabeele P, Donais MK. Mobile spectroscopic instrumentation in archaeometry research. App Spectrosc. 2016;70:27–41.
[65] Vandenabeele P, Edwards HGM, Jehlička J. The role of mobile instrumentation in novel applications of Raman spectroscopy: Archaeometry, geosciences, and forensics. Chem Soc Rev. 2014;43:2628–49.
[66] Vandenabeele P, Verpoort F, Moens L. Non-destructive analysis of paintings using Fourier transform Raman spectroscopy with fibre optics. J Raman Spectrosc. 2001;32:263–69.
[67] Vandenabeele P, Weis TL, Grant ER, Moens LJ. A new instrument adapted to in situ Raman analysis of objects of art. Anal Bioanal Chem. 2004;379:137–42.
[68] Vandenabeele P, Garcia-Moreno R, Mathis F, Leterme K, Van Elslande E, Hocquet F-P, et al. 'Multi-disciplinary investigation of the Tomb of Menna (TT69), Theban Necropolis, Egypt. Spectrochim Acta A. 2009;73:546–52.
[69] Deneckere A, Leeflang M, Bloem M, Chavannes-Mazel CA, Vekemans B, Vincze L, et al. The use of mobile Raman spectroscopy to compare three full-page miniatures from the Breviary of Arnold of Egmond. Spectrochim Acta A. 2011;83:194–99.
[70] Deneckere A, Schudel W, Van Bos M, Wouters H, Bergmans A, Vandenabeele P, et al. In situ investigations of vault paintings in the Antwerp Cathedral. Spectrochim Acta Part A. 2010;75:511–19.
[71] Vandenabeele P, Castro K, Hargreaves M, Moens L, Madariaga JM, Edwards HGM. Comparative study of mobile Raman instrumentation for art analysis. Anal Chim Acta. 2007;588:108–16.
[72] Jehlička J, Culka A, Vandenabeele P, Edwards HGM. Critical evaluation of a handheld Raman spectrometer with near infrared (785 nm) excitation for field identification of minerals. Spectrochim Acta Part A. 2011;80:36–40.

[73] Petrová Z, Jehlička J, Čapoun T, Hanus R, Trojek T, Goliáš V. Gemstones and noble metals adorning the sceptre of the faculty of science of Charles University in Prague: Integrated analysis by Raman and XRF handheld instruments. J Raman Spectrosc. 2012;43:1275–80.
[74] Conti C, Botteon A, Bertasa M, Colombo C, Realini M, Sali D. Portable sequentially shifted excitation Raman spectroscopy as an innovative tool for *in situ* chemical interrogation of painted surfaces. Analyst. 2016;141:4599–07.
[75] Jehlička J, Culka A, Košek F. Obtaining Raman spectra of minerals and carbonaceous matter using a portable sequentially shifted excitation Raman spectrometer – A few examples. J Raman Spectrosc. 2017;48:1583–89.
[76] Vagnini M, Gabrieli F, Daveri A, Sali D. Handheld new technology Raman and portable FT-IR spectrometers as complementary tools for the in situ identification of organic materials in modern art. Spectrochim Acta Part A. 2017;176:174–82.
[77] Tournié A, Prinsloo LC, Paris C, Colomban P, Smith B. The first *in situ* Raman spectroscopic study of San rock art in South Africa: Procedures and preliminary results. J Raman Spectrosc. 2011;42:399–06.
[78] Lahlil S, Lebon M, Beck L, Rousselière H, Vignaud C, Reiche I, et al. The first *in situ* micro-Raman spectroscopic analysis of prehistoric cave art of Rouffignac St-Cernin, France. J Raman Spectrosc. 2012;43:1637–43.
[79] Ravindran TR, Arora AK, Singh M, Ota SB. On-and off-site Raman study of rock-shelter paintings at world-heritage site of Bhimbetka. J Raman Spectrosc. 2013;44:108–13.
[80] Olivares M, Castro K, Corchón MS, Gárate D, Murelaga X, Sarmiento A, et al. Non-invasive portable instrumentation to study Palaeolithic rock paintings: The case of La Peña Cave in San Roman de Candamo (Asturias, Spain). J Archaeol Sci. 2013;40:1354–60.
[81] Hernanz A, Ruiz-López JF, Madariaga JM, Gavrilenko E, Maguregui M, Fdez-Ortiz De Vallejuelo S, et al. Spectroscopic characterisation of crusts interstratified with prehistoric paintings preserved in open-air rock art shelters. J Raman Spectrosc. 2014;45:1236–43.
[82] Pitarch A, Ruiz JF, Fdez-Ortiz De Vallejuelo S, Hernanz A, Maguregui M, Madariaga JM. *in situ* characterization by Raman and X-ray fluorescence spectroscopy of post-Paleolithic blackish pictographs exposed to the open air in Los Chaparros shelter (Albalate del Arzobispo, Teruel, Spain). Anal Methods. 2014;6:6641–50.
[83] Lauwers D, Candeias A, Coccato A, Mirao J, Vandenabeele P, Moens L. Evaluation of portable Raman spectroscopy and handheld X-ray fluorescence analysis (hXRF) for the direct analysis of glyptics. Spectrochim Acta A. 2016;15:146–52.
[84] Maguregui M, Knuutinen U, Martínez-Arkarazo I, Giakoumaki A, Castro K, Madariaga JM. Field Raman analysis to diagnose the conservation state of excavated walls and wall paintings in the archaeological site of Pompeii (Italy). J Raman Spectrosc. 2012;43:1747–53.
[85] Morillas H, Maguregui M, Gómez-Laserna O, Trebolazabala J, Madariaga JM. Characterisation and diagnosis of the conservation state of cementitious materials exposed to the open air in XIX century lighthouses located on the coast of the Basque country: The case of Igueldo lighthouse, San Sebastian, orth of Spain. J Raman Spectrosc. 2012;43:1630–36.
[86] Colomban P, Milande V. On-Site Raman analysis of the earliest known Meissen porcelain and stoneware. J Raman Spectrosc. 2006;37:606–13.
[87] Kırmızı B, Colomban P, Quette B. On-site analysis of Chinese cloisonne´ enamels from fifteenth to nineteenth centuries. J Raman Spectrosc. 2010;41:780–90.
[88] Lauwers D, Brondeel P, Moens L, Vandenabeele V. *in situ* Raman mapping of art objects. Philos Trans R Soc A. 2016;374:20160039.
[89] Realini M, Botteon A, Conti C, Colombo C, Matousek P. Development of portable defocusing micro-scale spatially offset Raman spectroscopy. Analyst. 2016;141:3012–19.

[90] Realini M, Conti C, Botteon A, Colombo C, Matousek P. Development of a full micro-scale spatially offset Raman spectroscopy prototype as a portable analytical tool. Analyst. 2017;142:351–55.
[91] Bikiaris D, Sister D, Sotiropoulou S, Katsimbiri O, Pavlidou E, Moutsatsou AP, et al. Ochre-differentiation through micro-Raman and micro-FTIR spectroscopies: Application on wall paintings at Meteora and Mount Athos, Greece. Spectroch Acta Part A. 1999;56:3–18.
[92] Ganitis V, Pavlidou E, Zorba F, Paraskevopoulos KM, Bikiaris D. A post-Byzantine icon of St Nicholas painted on a leather support. Microanalysis and characterisation of technique. J Cult Herit. 2004;5:349–60.
[93] Miliani C, Rosi F, Burnstock A, Brunetti BG, Sgamellotti A. Non-invasive in-situ investigations versus micro-sampling: A comparative study on a Renoirs painting. Appl Phys A. 2007;89:849–56.
[94] Anastasiou M, Hasapis T, Zorba T, Pavlidou E, Chrissafis K. Paraskevopoulos KM. TG-DTA and FTIR analyses of plasters from byzantine monuments in Balkan region, Comparative study. J Therm Anal Cal. 2006;84:27–32.
[95] Nakai I, Abe Y. Portable X-ray powder diffractometer for the analysis of art and archaeological materials. Appl Phys A. 2012;106:279–93.
[96] Van De Voorde L, Van Pevenage J, De Langhe K, De Wolf R, Vekemans B, Vincze L, et al. Non-destructive in situ study of "Mad Meg" by Pieter Bruegel the Elder using mobile X-ray fluorescence, X-ray diffraction and Raman spectrometers. Spectrochim Acta Part B. 2014;97:1–6.
[97] Veneranda M, Costantini I, Fdez-Ortiz De Vallejuelo S, Garcia L, García I, Castro K, et al. Study of corrosion in archaeological gilded irons by Raman imaging and a coupled scanning electron microscope–Raman system. Philos Trans A Math Phys Eng Sci. 2016;374:20160046.
[98] Deneckere A, Vekemans B, Van De Voorde L, De Paepe P, Vincze L, Moens L, et al. Feasibility study of the application of micro-Raman imaging as complement to micro-XRF imaging. Appl Phys A. 2012;106:363–76.
[99] Van De Voorde L, Vandevijvere M, Vekemans B, Van Pevenage J, Caen J, Vandenabeele P, et al. Study of a unique sixteenth century Antwerp majolica floor in the Rameyenhof castle's chapel by means of X-ray fluorescence and portable Raman analytical instrumentation. Spectroch Acta Part B. 2014;102:28–35.
[100] Wehling B, Vandenabeele P, Moens L, Klockenkiimper R, Von Bohlen A, Van Hooydonk G, et al. Investigation of pigments in medieval manuscripts by micro Raman spectroscopy and total reflection X-yay fluorescence spectrometry. In: Mikrochim Acta. vol. 130. 1999. p. 253–60.
[101] Beck L, Rousseliere H, Castaing J, Duran A, Lebon M, Moignard B, et al. First use of portable system coupling X-ray diffraction and X-ray fluorescence for *in-situ* analysis of prehistoric rock art. Talanta. 2014;129:459–64.
[102] Andrikopoulos KS, Daniilia S, Roussel B, Janssens K. *In vitro* validation of a mobile Raman–XRF micro-analytical instrument's capabilities on the diagnosis of Byzantine icons. J Raman Spectrosc. 2006;37:1026–34.
[103] Van Der Snickt G, Dubois H, Sanyova J, Legrand S, Coudray A, Glaude C, et al. Large-area elemental imaging reveals Van Eyck's original paint layers on the Ghent Altarpiece (1432), rescoping its conservation treatment. Angew Chem. 2017;129:4875–79.

Marisia A. Fikiet, Shelby R. Khandasammy, Ewelina Mistek, Yasmine Ahmed, Lenka Halámková, Justin Bueno and Igor K. Lednev

9 Forensics: evidence examination via Raman spectroscopy

Abstract: Forensic science can be broadly defined as the application of any of the scientific method to solving a crime. Within forensic science there are many different disciplines, however, for the majority of them, five main concepts shape the nature of forensic examination: transfer, identification, classification/individualization, association, and reconstruction. The concepts of identification, classification/individualization, and association rely greatly on analytical chemistry techniques. It is, therefore, no stretch to see how one of the rising stars of analytical chemistry techniques, Raman spectroscopy, could be of use. Raman spectroscopy is known for needing a small amount of sample, being non-destructive, and very substance specific, all of which make it ideal for analyzing crime scene evidence. The purpose of this chapter is to show the state of new methods development for forensic applications based on Raman spectroscopy published between 2015 and 2017.

Keywords: forensics, criminalistics, trace evidence, bloodstain, biological stain, gunshot residue, chemometrics, counterfeit pharmaceuticals, controlled substance, toxicology, explosive, questioned document, trace analysis, fiber, hair

9.1 Introduction

Forensic science can be broadly defined as the application of any part of the scientific method to solving a crime. Within forensic science there are many different disciplines. For the majority of them, however, five main concepts shape the nature of forensic examination: transfer, identification, classification/individualization, association, and reconstruction [1]. The concepts of identification, classification/individualization, and association rely greatly on analytical chemistry techniques. It is, therefore, no stretch to see how one of the rising stars of analytical chemistry techniques, Raman spectroscopy, could be of use. Is this paint chip found on the victim from the paint on the killer's car? Is this white powder cocaine or flour? Raman spectroscopy is known for

This article has previously been published in the journal *Physical Sciences Reviews*. Please cite as: Fikiet, M. A., Khandasammy, S.R., Mistek, E., Ahmed, Y., Halámková, Bueno, J., Lednev, I. K. Forensics: evidence examination via Ramanspectroscopy *Physical Sciences Reviews* [Online] **2019**, 4. DOI: 10.1515/psr-2017-0049.

https://doi.org/10.1515/9783110515312-009

needing a small amount of sample, being non-destructive, and very substance specific, all of which make it ideal for analyzing crime scene evidence.

In recent years there has been a push for increased scrutiny of the validity of currently used forensic techniques [2]. This has led to an increase in research into forensic science methods, and into new methods that could be used in the future. Raman spectroscopy has been used in the majority of the different disciplines and has imparted its many advantages to the study of forensic science. Chemometrics has also been added in many studies to give mathematical certainty and quantitative predictors to various types of evidence.

The purpose of this article is to show the state of new methods development for forensic applications based on Raman spectroscopy since reviews by Doty et al. [3] and Muro et al. [4] and to expand on the newly published mini review by Doty et al. [5]. The majority of discussed material was published between 2015 and 2017. Additional resources about presented forensic discipline are given in the beginning of the each section.

9.2 Chemometrics

Raman spectroscopy has consistently confirmed its importance in many areas, all due to its nondestructive nature, speed, and low cost. However, a small complication is that Raman spectroscopy results in a large amount of high-dimensional data. This data requires advanced statistical analysis to properly understand and utilize the recovered information. The rapid development of innovative methods and significant technological advances have helped to increase the interest in Raman spectroscopy. For a more in-depth treatment of chemometrics the reader is referred to Chapter 4 of this text.

Principal component analysis (PCA) is one of the most popular chemometric techniques. Salahioglu et al. [6] performed PCA prior to the k-nearest neighbors (KNN) analysis in order to reduce dimensionality of the dataset to classify lipstick samples. A 98.7 % correct classification rate was obtained when data was acquired from glass slides, and 100 % classification rate once the lipsticks were transferred onto different fibers. PCA can further be used to separate data based on similarities between the samples. Such an example can be found in the study carried out in 2017 by Mohamad Asri et al. [7]. Researchers were able to group together different colors of ballpoint pen ink using PCA analysis. Discriminant analysis, on the other hand, predicts the classification of unknown samples into one of the known classes. Mistek et al. [8] implemented support vector machines discriminant analysis (SVM-DA) with Raman spectroscopy in order to differentiate donor race based on the analysis of blood samples. In a study performed by Muro et al. [9], both SVM-DA and partial least squares discriminate analysis (PLS-DA) models were evaluated for their ability to differentiate five different body fluids. In this study, the models used either the entire spectral range or variable selection. In contrast to discriminant analysis, regression models predict quantitative values. Doty et al. [10, 11] investigated ageing of bloodstains in detail using regression models. They trained partial least squares regression (PLSR) models to relate spectral intensities with time since deposition (TSD) of the blood sample. The two studies investigated TSD up to 1 week [10] and 2 years [11].

9.3 Biological samples

Biological samples are of high importance during forensic investigation. They can be used to connect a crime to an individual (such as a perpetrator or a victim). Current techniques for body fluid analysis are still not universal; meaning a single test identifies only a single body fluid [12]. Presumptive and confirmatory tests, which are specific for different body fluids, can still result in false-positives (another substance is misidentified as a particular body fluid). Moreover, most of these tests are destructive to the sample due to the reagents used to initiate the reaction.

Laser based spectroscopic techniques, including Raman spectroscopy, have been promising for applications in forensic examinations [13]. The application of Raman spectroscopy for analysis of body fluids was widely studied by the Lednev research group. Their first publication in 2008, explained the potential of this approach [14]. The Raman spectroscopic signatures of the main body fluids, semen, blood, saliva, sweat, and vaginal fluid, were explored using Raman spectroscopic mapping combined with statistical analysis [15–19]. Blood traces were one of the main focuses in analysis of body fluids. Raman spectroscopy combined with chemometrics was successful in the discrimination of menstrual blood from peripheral blood [20]. Moreover, the identification of species from a bloodstain was also achieved using this technique [21]. Body fluids found at crime scenes are very often deposited on different materials. Therefore, the study of blood deposited on various substrates was performed [22]. Additionally, the effect of laser power on a bloodstain was studied [23]. The Lednev group also published review articles [3, 4] showing the development of vibrational spectroscopy for forensic purposes, including body fluids. Since the aforementioned review articles published by Muro et al. and Doty et al., new studies have proven the potential of Raman microspectroscopy for body fluids examination.

A new approach for discriminating between different body fluids was proposed by Muro et al. [9]. The identification of biological stains was achieved using chemometrics to build classification models. The study included the discrimination between the following human body fluids: peripheral blood, saliva, semen, sweat, and vaginal fluid. Different modeling techniques were applied, and the best performance was achieved using SVM-DA which showed 100 % accuracy for external validation using new samples. Semen and blood specifically were studied by Zou et al. [24] using infrared and Raman spectroscopy. In their study, these techniques showed nondestructive, confirmatory identification of the two body fluids. Feine et al. [25], combined the prostate specific antigen (PSA) test and micro-Raman spectroscopy in order to detect semen. The PSA test alone is not confirmatory since it can give false positive results. Urine is one of the substances that can give a false positive with the PSA test. In this study, Raman spectroscopy was found to be suitable for nondestructive differentiation between semen and urine.

Blood was studied not only for forensic purposes, but also for medically-related applications. Recently, Atkins et al. [26] published a review article on Raman spectroscopic analysis of blood and its components. During the forensic blood testing procedures, species identification may be one of the steps performed [12]. However, currently used assays species identification are destructive to the sample. Therefore, a method which could preserve the sample is ideal. Bai et al. [27] reported on a study which used Raman spectroscopy with modeling to differentiate between human and nonhuman blood and showed great accuracy for discrimination.

Besides identification of the origin of a biological stain, the TSD is another important aspect for forensic analyses. Doty et al. studied the TSD of a bloodstain for stains up to one week [10], and even up to two years [11]. To use Raman spectroscopy to study a bloodstain up to one week after deposition, researchers used two-dimensional correlation spectroscopy (2D CoS) and statistical modeling (Figure 9.1) [10]. The 2D CoS and PLSR modeling allowed for analysis of kinetics effects of the bloodstain aging and prediction of the time since the stains' deposition. In the second

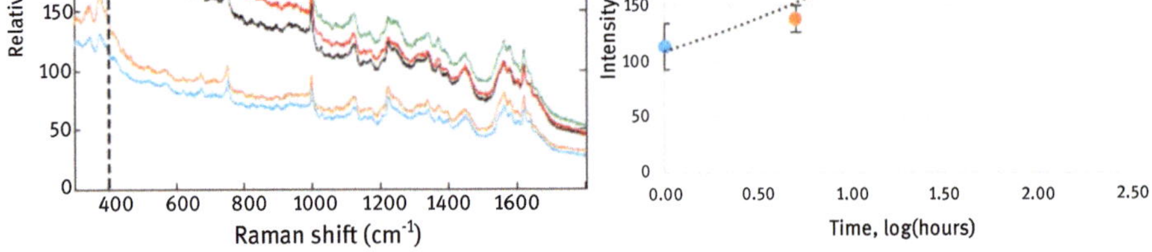

Figure 9.1: (a) The averaged raw spectra at increasing time points that show the change in fluorescence and (**b**) the trend of the intensity of the peak at 400.5 cm^{-1} for all raw spectra over time in the exponential fit trend line. Reprinted from Doty KC, McLaughlin G, Lednev IK. A Raman "Spectroscopic clock" for bloodstain age determination: the first week after deposition, Analytical and Bioanalytical Chemistry 408 (2016) 3993–4001. (ref. no.10) Copyright 2016 with permission from Springer.

study [11], two different regression models were used, PLSR and principal component regression (PCR). Raman spectra of bloodstains were also tested with the developed methodology for body fluid identification [9] in order to estimate whether the spectra of aged blood are still predicted as peripheral blood in the model containing main body fluids (i. e., blood, saliva, semen, sweat, and vaginal fluid). It was shown that all spectra of blood aged up to 1 month were identified as blood and not another body fluid.

One of the problems associated with crime scenes is that the sample size can be limited. Therefore, the ideal method for analysis and characterization of body fluids found at a scene should be sensitive and nondestructive to the sample. Muro and Lednev [28] studied the limit of detection of Raman spectroscopy using blood samples. They found that a single red blood cell is enough for the sample to be classified as peripheral blood based on its Raman spectrum. For comparison, 50 pg of DNA are required to obtain a DNA profile, which corresponds to around 1250 pL of blood. Using Raman spectroscopy the detection of the bloodstain was possible with only 250 fL.

Furthermore, Raman spectroscopy was used to obtain information about fluid donors based on traces of different body fluids. In two different studies blood samples were used to differentiate between human sexes [29] and races [8]. Since the variations in blood samples between different donors come only from the quantitative, and not qualitative, changes in their composition, advanced statistical analysis was a necessary step. Sikirzhytskaya et al. [29] used SVM and artificial neuron network (ANN) methods in order to achieve discrimination between the sexes of human donors. The best findings were achieved using a combination of genetic algorithm (GA) with an ANN classification, resulting in approximately 98 % accuracy at the bloodstain level. The study on race differentiation performed by Mistek et al. [8] focused on discrimination between Caucasian and African American blood donors. They built SVM-DA models which were evaluated using receiver operating characteristic (ROC) with area under the curve (AUC) analysis. Using this method researches obtained results of 83 % probability of proper classification of an individual donor based on outer loop subject-wise cross-validation.

The same research group utilized Raman microspectroscopy to obtain details about individuals using body fluids other than blood. Saliva samples were used to determine the sex of human donors [30], and semen samples were used to determine the donors race [31]. A SVM-DA model and ROC curve were used to classify saliva by sex, resulting in 94 % accuracy for calibration donors and 92 % accuracy using external donors [30]. In the study on race differentiation from semen samples [31], GA was used to select variables showing the largest differences between Caucasian and African American donors. Based on SVM-DA and ROC curve, the approach showed 100 % correct classification to the proper race at the donor level; all donors were predicted correctly (18 donors from calibration and 7 donors from external validation).

9.4 Controlled substances

One of the foremost problems facing society today is worldwide drug abuse, which is correlated to an increase in drug-related health problems [32]. The United States Drug Enforcement Administration typically separates drugs into the following categories: depressants (e. g. barbiturates, benzodiazepines), designer drugs (e. g. bath salts, cathinones, spice), hallucinogens (e. g. ketamine, psilocybin/mushrooms, peyote/mescaline), inhalants (marijuana/cannabis), narcotics (e. g. heroin, methadone, oxycodone, fentanyl), steroids, and stimulants (e. g. amphetamine, methamphetamine, cocaine) [33]. Not all drugs that are abused are illegal. Prescription drugs such as benzodiazepines (alprazolam/Xanax®, clonazepam/Klonopin®, diazepam/Valium®) and prescription opioids (morphine, oxycodone/OxyContin®) are commonly sold and abused [33]. The increased in "legal highs" (legal substances that produce effects similar to those produced by illicit drugs) is also a growing concern [34].

Drugs are often mixed with other substances (adulterants) to both increase a drug dealer's profits and hide the actual drug from law enforcement agencies [35]. Since many of these unknown additives and fillers can be hazardous and, in some cases, potentially lethal, quick and accurate identification of the components of an unknown drug sample is crucial [35]. Identification of the active agent usually starts with preliminary color tests that can help sort the substance into the different classes mentioned previously [36]. Thin layer chromatography (TLC) is also a common method for screening samples before confirmatory analysis. Confirmatory analysis is usually carried out using a variety of analytical chemistry techniques such as gas chromatography coupled with mass spectrometry (GC-MS), high performance liquid chromatography (HPLC), capillary electrophoresis (CE) and infrared spectrometry (IR)[36].

Rebiere et al. analyzed eight samples of anabolic tablets using Raman microscopy in conjunction with near-IR spectroscopy [37]. Anabolic tablets are frequently and illegally used by athletes due to the drugs' performance-enhancing properties. In analyzing various samples of these drugs, the authors found that several of the samples contained steroids, as well as lactose, sucrose, starch, and talc. Research groups such as Jones et al. and Elie et al. found that Raman microscopy could also be used to identify psychoactive drugs when combined with other tests and instrumentation [38, 39]. Jones et al. was able to correctly identify psychoactive drugs when using Raman microscopy combined with IR spectroscopy [38]. Once these researchers extended the spectral libraries of these instruments, they were able to identify 76 % of the 221 psychoactive substances they analyzed in their study. The combination of Raman microscopy and microcrystalline tests allowed Elie et al. to identify two novel psychoactive substances—5,6-methylenedioxy-2-aminoindane (MDAI) and 4-methylmethcathinone (mephedrone)—despite the presence of other cutting agents and species [39].

9.4.1 Counterfeit pharmaceuticals

Counterfeit pharmaceuticals represent an increasing risk to public health. Contributing factors to the surge of counterfeit pharmaceuticals include increasing internet sales, counterfeiting sophistication and demand for treatments [40, 41]. Counterfeits may take several forms including deceptive, adulterated, mislabeled and substandard medicines. Deceptive or fraudulent products often possess no active pharmaceutical ingredient (API), the wrong API or the wrong amount of the API. Substandard and adulterated medicines are often not subjected to the correct quality control, may be mislabeled, and/or contain toxic impurities.

For investigators, key analytical markers must be identified to determine if a sample is a counterfeit. Identifying and quantitating the API is an important step for detecting counterfeit pharmaceuticals. Historically, rapid and inexpensive techniques such as TLC have been used for counterfeit detection. Modern kits, such as the TLC "mini-lab", are very useful for quick on-site analysis [42]. Although TLC is a rapid screening method, it is nonspecific and requires further analysis for conformation. Chromatographic methods, such as HPLC, are used for quantitative API determinations. HPLC offers high sensitivity but unfortunately requires time consuming sample preparation, analysis time, and method development. Raman micro-spectroscopy possesses inherent selectivity and rapid analysis capabilities with minimal sample preparation, which offers improvements over current methods for counterfeit detection. Unfortunately, Raman spectroscopy for counterfeit detection has historically suffered from interferences form inactive ingredients (excipients) which can mask signal from the informative API. Detecting these excipients is not an indication that the sample is a genuine pharmaceutical, as counterfeits often have the same inactive ingredient matrixes as genuine pharmaceuticals. Raman spectroscopic analysis also has had trouble with sensitivity, especially with low API dosage forms. Reviews by Aboul-Enein et al. and Muro et al. described previous research utilizing Raman spectroscopy and portable Raman spectroscopy for anti-counterfeiting [4, 43]. In this review, advances in Raman micro-spectroscopy for the analysis of counterfeit pharmaceuticals are emphasized.

Loethen and co-workers targeted counterfeit identification through the use of Raman spectroscopic data and a spectral correlation (SC) multivariate analysis approach [44]. Reference spectra collected from the isolated API of different formulations were compared to Raman spectra from finished drug products and simulated counterfeits. It was found that optimizing the spectral range to not select regions with peaks from the excipients, dyes and coatings were found to improve the spectral correlation between the API library and the Raman spectra from the samples, maximizing true positive assignments. The technique was able to correctly identify five counterfeit samples which were void of the API.

Packaging of genuine and counterfeit pharmaceuticals was compared with 2D CoS for the drug product Cialis® [45]. Counterfeit Cialis® was obtained from the internet and the packaging between the genuine and fake samples were compared. The approach was capable of identifying differences in the chemical composition of the packaging even within white colored regions which appeared similar to the naked eye. The confocal Raman spectroscopic approach combined with the 2D CoS analysis was found to be capable of providing a visual map of the spectroscopic space and visually illustrating different chemical components.

9.4.2 Toxicology

Toxicology in this instance refers to identification of a substance once it has already been consumed, and therefore the sample type is usually a body fluid containing the controlled substance or its metabolites [46]. Because of the sample matrix, screening methods vary from traditional controlled substance methods. Following collection, a sample is analyzed using presumptive tests, such as enzymatic immunoassays, to determine if drugs are present in the sample [47]. If presumptive tests give results that suggest the presence of one or several drugs, the sample is then analyzed using confirmatory tests [47]. Confirmatory testing typically involves the use of a combination of analytical methods such as HPLC, GC-MS, IR, UV spectrophotometry, and X-ray diffraction [35, 47, 48]. While these techniques are incredibly useful for detecting drugs, they are problematic in that they are expensive and time-consuming; furthermore, these techniques typically involve destruction of the sample and the use of dangerous solvents [35, 48].

D'Elia et al. used Raman spectroscopy with a 532 nm excitation to analyze both street cocaine samples and samples of nasal fluid spiked with cocaine [49]. Using this method, the researchers were able to detect cocaine in nasal fluid down to a concentration of approximately 200 µg, regardless of whether the nasal fluid samples were fresh and wet or old and dry (Figure 9.2). The authors noted that even though an extraction is necessary prior to sample analysis, Raman microscopy is invaluable in the field of drug detection.

9.5 Explosives and gunshot residue

Explosives evidence is different from other types of trace evidence in that it has the potential to cause direct physical harm [50]. The fact that explosive compounds can decompose swiftly drives their destructive power [51]. Homeland security groups and law enforcement agencies respectively concern themselves with pre and post blast inquiries [51]. Explosives can be classified a number of different ways [50].

Forensic laboratories utilize a number of analysis methods to detect explosive compounds; these include chromatography, mass spectrometry, Infrared spectroscopy, and HPLC [50]. Several factors present Raman spectroscopy as a viable

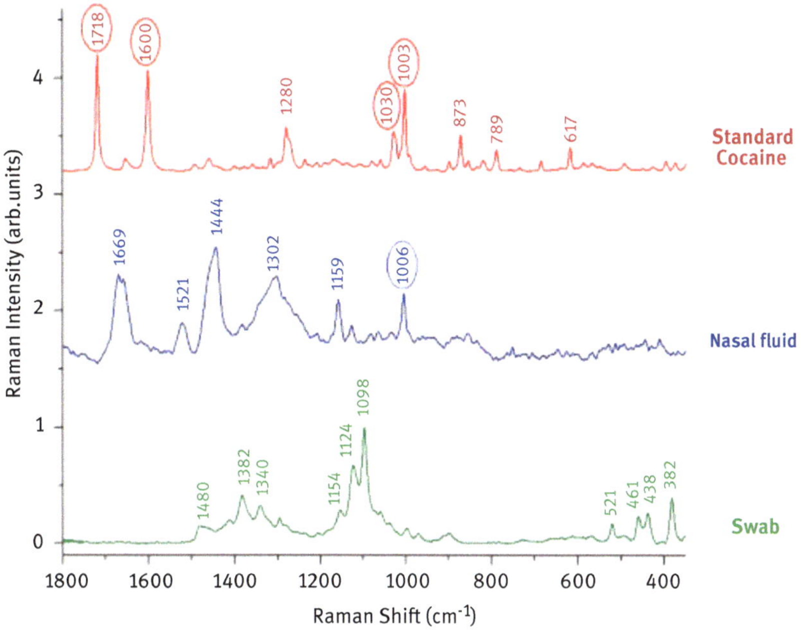

Figure 9.2: Spectra of standard cocaine, nasal fluid and swab (SWAB1). All the important bands have been indicated. The ones circled in the first spectrum correspond to the characteristic bands of cocaine, the one in the second spectrum corresponds to the possible interfering band of the nasal fluid with the cocaine. Reprinted from D'Elia V, Montalvo G, Ruiz CG, Analysis of street cocaine samples in nasal fluid by Raman spectroscopy, Talanta 154 (2016)367–73. (ref. no. 49) Copyright 2016 with permission from Elsevier.

technique for explosives analysis, but most notably, Raman spectroscopy is a portable technique and is nondestructive [51, 52].

9.5.1 Explosives

In 2016, three papers were released regarding the application of Raman spectroscopic techniques for the investigation of IEDS [52]. IEDS or Improvised Explosive Devices are homemade devices created for the purpose of causing destruction and have gained relevance in current times due to their association with acts of terrorism [52]. A common IED, erythritol tetranitrate, can be made from low calorie sweetener. Matyáš and coauthors found through investigation of this compound that the Raman spectrum contained peaks stemming from NO_2 symmetrical stretching and N-O stretching [53]. Zapata and coauthors published two manuscripts involving the analysis of explosives using Raman spectroscopy. One of these manuscripts investigated the analysis of trace compounds stemming from IEDS found in handprints (Figure 9.3)[52]. Researchers

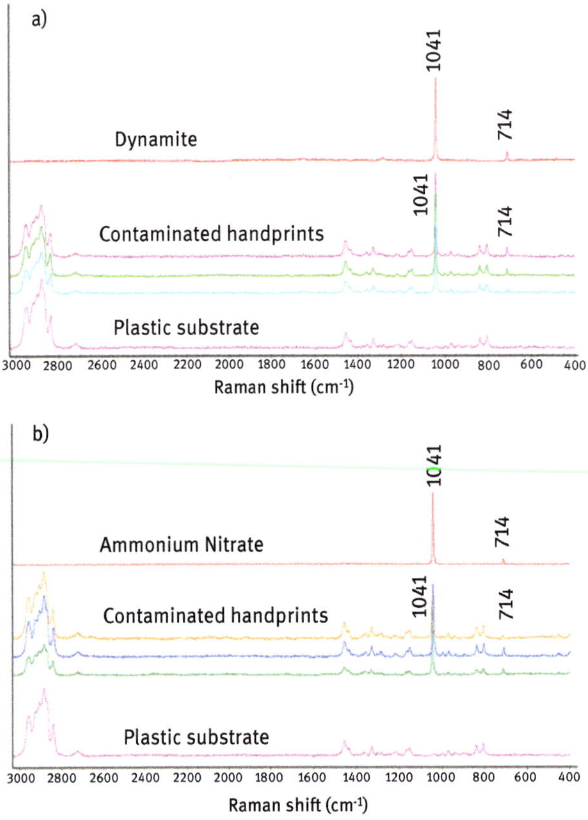

Figure 9.3: Dynamite and ammonium nitrate identification in handprints. Spectral comparisoin of (a) standard dynamite, three different handprints from donors after handling dynamite and the substrate and (b) standard ammonium nitrate, handprints from three donors after handling ammonium nitrate and the substrate. Reprinted from Zapata F, Fernández De La Ossa Á, Gilchrist E, Barron L, García-Ruiz C. Progressing the analysis of Improvised Explosive Devices: Comparative study for trace detection of explosive residues in handprints by Raman spectroscopy and liquid chromatography, Talanta 161 (2016), 219–27. (ref. no.52) Copyright 2016 with permission from Elsevier.

achieved this by comparing Raman spectroscopic and liquid chromatography analyses. The substances subjected to analysis included black powder, ammonium nitrate, dynamite, and smokeless powders. It concluded that Raman spectroscopy was useful for the detection of black powder, ammonium nitrate, and dynamite. However, it was not possible to distinguish smokeless powders using Raman spectroscopy. Liquid chromatography was deemed to provide a complement for the Raman spectroscopic analyses that were carried out.

In their second manuscript, Zapata et al. investigated particles that formed as the result of ten open air explosions resulting from ten distinct explosive compounds using Raman spectroscopy [54]. The explosive compounds that were selected are all common in IEDS and included: black powder, trinitrotoluene (TNT), dynamite, ammonal, hexamethylene triperoxide diamine (HMTD), chloratite, RDX, pentaerythritol tetranitrate (PETN), ammonium nitrate fuel oil (ANFO), and triacetone triperoxide (TATP). The Zapata group was not able to detect particles that had not reacted post blast when examining pure compounds. The researchers were successful in identifying all postblast particles resulting from mixtures that were based upon oxidizing salts. The group also reported on the ability of their analysis to discriminate between potassium nitrate and ammonium nitrate without any preparation of the samples. This was considered to be rather notable as other types of analysis require sample preparation involving the dissociation of these salts.

The analysis of explosives using Raman spectroscopy is very relevant because of the impact explosives have on public safety and the role they play in current events. Explosives are varied in composition and presentation and Raman spectroscopy has presented itself as viable tool to meet the challenges of explosives analysis.

9.5.2 Gunshot residue

Gunshot residue (GSR) is comprised of tiny particles that are produced when a firearm is discharged [55]. There are two types of GSR, inorganic and organic GSR, which can be differentiated based on size and chemical composition [56]. Typical GSR analysis is dependent upon the detection of the three elements Pb, Sr, and Ba with scanning electron microscopy coupled with energy dispersive x-ray spectroscopy (SEM- EDX) [56, 57]. With the advent of new eco-friendly cartridges that lack lead, SEM- EDX cannot be so deeply replied upon [56].

The Lednev research lab, made some important progress investigating using Raman spectroscopy to analyze GSR [55]. In a series of papers published from 2012 to 2014, several important studies occurred. Bueno et al. investigated the relationship between ammunition caliber, manufacturer, and GSR using Raman spectroscopy (Figure 9.4) [55], as well as using FT-IR analysis to improve GSR discrimination [57]. Lastly, a method was published on using Raman spectroscopic mapping for GSR particles on tape [56].

The López-López research group published a notable paper on the application of Raman spectroscopy for the analysis of GSR. In 2015, López-López and coauthors used Raman spectroscopy to perform fast mapping of GSR particles on various substrates [58]. The overall goal was to create a method to quickly analyze GSR particles. The GSR from different cartridges was analyzed with an excitation wavelength of 455 nm on a variety of different substrates, and it was concluded that Raman spectroscopy has the potential to be used for GSR mapping on various substrates.

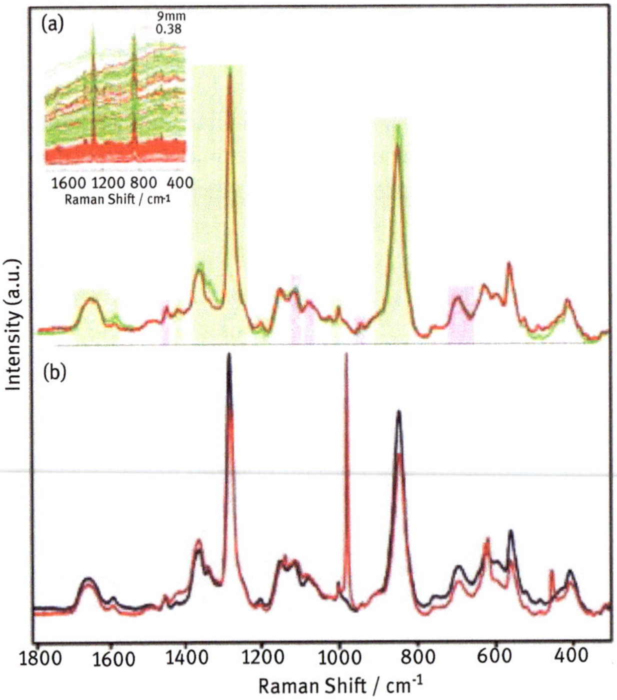

Figure 9.4: (a) The selected Raman spectra collected from 9 mm (red line) and 0.38 (green line) caliber GSR of different caliber GSR particles. Shaded areas indicate the contributions of the inorganic (purple) and organic (green) substituent's from the original ammunition. (b) Illustration of the heterogeneity of the GSR as the particle level. The red and blue spectra were collected from different locations on the same GSR particle. Reprinted from Bueno J, Sikirzhytski V, Lednev IK, Raman Spectroscopic Analysis of Gunshot Residue Offering Great Potential for Caliber Differentiation, Analytical Chemistry 84 (2012) 4334–9. (ref. no.55) Copyright 2012 with permission from the American Chemical Society.

9.6 Questioned documents

The examination of questioned documents requires analysis to provide information about the writer's handwriting and signature, as well as information about the ink and paper the document was written on. This can include the origin, history, and dating of a document as well as any changes made (e.g., alterations, deletions, and additions to documents) [59]. When looking at the writer's handwriting, document examiners look for both class and individual characteristics [60]. Class characteristics refer to the style, since different styles of writing are taught in different countries and regions. Styles can also change with time. Unusual characteristics of handwriting are another type of feature that can often be useful. However, these attributes are not enough to individualize handwriting. Furthermore, determining when a document was written is still an ongoing problem [61]. One of the ways to attempt this is the analysis of office tools and technologies used on the document in question and determining their dates of invention. Moreover, features that may appear to be from a typewriter and/or photocopier can be analyzed, as well as supplies used in the production of the paper/inks. Aging of paper/ink can be also used during examination of questioned documents. Analytical techniques that are commonly used in the examination of ink and paper include chemical, optical, spectroscopic, and chromatographic methods [62].

Questioned documents are one type of evidence that can be found at a crime scene. New research is still being performed to find better and more reliable methods for the examination of questioned documents [4, 63, 64]. Different techniques applied in analysis of questioned documents were reviewed by Calcerrada and García-Ruiz [63]. They covered publications between 2000 and 2014 and noticed increasing attention to spectroscopic techniques over separation techniques. Moreover, almost 50 % of studies utilizing spectroscopic techniques have employed analytical mass spectrometry. The remaining papers were based on techniques like infrared, Raman or UV–vis spectroscopy. For more information on Raman spectroscopy and document examination, the reader is referred to the following reviews. Muro et al. [4] reviewed the application of vibrational spectroscopy for analyzing different types of forensic evidences. Concentration on Raman spectroscopy for analysis of pigments and dyes in ink and paint was reported by Buzzini and Suzuki [64]. Since the previous reviews, new studies have already been performed and published showing potential of Raman microspectroscopy for the analysis of questioned documents.

Ink analysis is one of the most widely studied aspects of document examination for forensic purposes. The interest was shown in previous years, as can be seen in the aforementioned articles, and continues today. Different analytical techniques, including Raman spectroscopy, were performed on pen ink by Nunkoo et al. [65]. However, some inks exhibited fluorescence, making it impossible to collect Raman spectra. The nondestructive nature of Raman spectroscopy was pointed out by Lee et al. [66]. They found Raman spectroscopy to be a rapid and practical tool for ink analysis. In their study, Raman spectroscopy was utilized to discriminate blue and black ballpoint pen inks. The classification was based on the total number of prominent bands in Raman spectra of inks. Overall, 94 % and 95 % of samples were successfully classified as blue and black inks, respectively.

Raman spectroscopy was combined with chemometrics by Mohamad Asri et al. [7, 67]. A PCA model improved the classification performance of different groups of red and blue ballpoint pen inks [7]. In their other study [67], Pearson's product moment correlation (PPMC) coefficients and PCA were utilized to examine Raman and infrared spectra of red gel pen inks. Another chemometric approach, multivariate curve resolution-alternating least squares (MCR-ALS), was used by Borba et al. [68]. In their study MCR-ALS was applied to study cross lines and obliteration cases using confocal Raman spectroscopic imaging. The investigation of falsified documents was also studied by Huynh et al. [69]. They combined different techniques, including Raman spectroscopy, with direct analyte-probed nanoextraction (DAPNe). Raman spectroscopy was used to detect chemically-active components of ink, and showed a lot of advantages since it is nondestructive and does not require sample preparation. However, the combination of DAPNe with Raman spectroscopy was not able to distinguish altered parts of writing when the same pen was used.

Besides pen ink analysis, printing inks were recently examined by Johnson et al. [70] using Raman microspectroscopy. Two different excitation wavelengths were used, 532 and 785 nm, and the 785 nm laser was found to be more suitable for application in printing ink analysis. Moreover, Chen et al. [71] utilized confocal Raman spectroscopy for discrimination of inks used to create seals on documents. The method allowed for detection of some pigments present in the seal ink samples.

In addition to the analysis of different types of inks, Raman spectroscopy was found suitable for paper analysis as well [72, 73]. Alves et al. [72] performed a study on Japanese, paperboard, filter, filiset, and siliconed papers. Cotton and wood standards were used in order to compare the Raman signal and evaluate the structural properties of cellulosic materials. The ageing and degradation process of paper were studied by Zięba-Palus et al. [73] using Raman and infrared spectroscopy. When the two techniques were compared, infrared spectroscopy provides greater sensitivity and is free from fluorescence effect.

9.7 Trace analysis

Trace or transfer evidence covers easily transferable materials that do not fit into other evidence categories, such as glass, soil, hair, paint, dust, pollen, and fibers [74]. Paint, hair, and fibers are some of the most common types of trace evidence and are the types discussed here. Since all of these materials greatly differ in composition, analysis methods differ as well. However, analysis of all of these materials starts with microscopy. This type of evidence is often used to associate different parts of a crime, such as the perpetrator with the victim, or a body with a car used to carry it to a dump site [74].

9.7.1 Paint

The definition for paint is wide-ranging and can be used to describe any material whose purpose is to cover a variety of surfaces for protection or decoration [75]. The three major components of paint are the binder, pigment, and solvent. The solvent, which evaporates during the drying process, is not used in forensic identification. Forensic identification is, therefore, mostly focused on the identity of the pigment, binder, and other additives [76]. Visual examination using a microscope to determine the number, order, and color of layers contained in the paint sample is the first and most important part of forensic paint analysis [76]. Information about the binder is usually found using Fourier transform infrared spectroscopy (FT-IR) and pyrolysis gas chromatography (GC), while Raman spectroscopy has been more commonly used in recent years to identify pigments [76, 77]. For this review, the two major subsets of forensic paint analysis that will be covered are automotive and artistic paints. Artistic paint samples are used in forgery identification, while automotive paints are used mostly in automobile accidents and hit and runs [64]. For more information on forensic paint analysis, the reader is referred to *Forensic Examination of Glass and Paint: Analysis and Interpretation* [78].

9.7.1.1 Automotive

Ferreira et al. [79] studied the optimum conditions for performing Raman spectroscopy of automotive paint samples. Thirty-six paint samples from different manufacturers and in a variety of colors were studied. These samples were used to evaluate 10 different parameters, including laser power, excitation wavelength, microscope objective, and sample preparation. Experimentation showed that a 785 nm excitation worked best with a 50× objective and a high laser power. Interestingly, they found that spectra taken from paint coated over plastic had more peaks than spectra taken from paint on metal. Additionally, they found that taking spectra from the top of the paint chip, and not from a cross section, increased reproducibility of the spectra.

Chen et al. [80] investigated an 11 layer paint chip using a variety of techniques, including Raman spectroscopy. Raman spectroscopy, with an excitation wavelength of 785 nm, was used to identify the different pigments and additives that made up the various layers. High fluorescence of some layers prevented them from being identified with Raman spectroscopy. The pigments phthalocyanine blue and iron (III) oxide, as well as the additives talc and dolomite, were identified with Raman spectroscopy. All 11 layers were able to be characterized when the other techniques were used in conjunction with Raman spectroscopy.

Lv et al. combined information from both Raman spectroscopy and IR in two published studies. In the first study [81], IR was used to discriminate resin types, while Raman spectroscopy was used to discriminate similar colors of paint. Three green paint chips, two red paint chips, two orange paint chips, and two blue paint chips were analyzed with a Raman microscope using a 785 nm laser. All groups of paint chips could be differentiated when both IR and Raman spectroscopy were used. The second study [82] used Raman spectroscopy to identify the presence of lead chromate in Chinese car paint. Lead chromate is of particular interest because it has been banned in the USA for use in paint due to environmental reasons, but is still used in China. Of the 52 car paint samples analyzed, 12 of them contained lead chromate. It is interesting to note that Raman spectroscopy was also able to differentiate two different types of titanium dioxide present in the paint. Raman spectroscopy was able to improve discrimination power for all paint samples.

Martyna et al. [83] and Michalska et al. [84] used Raman spectroscopy and chemometrics to help distinguish blue car paints. Martyna et al. used wavelet transformations and models based on likelihood ratios to differentiate 25 solid and 30 metallic blue car paints with a 785 nm laser. Wavelet transformation was used for baseline correction and was able to preserve the chemical information in the spectra. After the transformation, univariate and multivariate models were made for both solid and metallic paints. It was found that multivariate models were able to distinguish the paint samples better than univariate models. Michalska et al. used a similar method and model type but were able to decrease the amount of variables used while still preserving the discrimination power.

Zieba-Palus et al. [85] improved the resulting spectra of blue car paint using photo bleaching. Twenty blue cars samples, both solid and metallic, were used. Prior to photo bleaching, peaks were evident in only 3 out of the 20 samples. A variety of different bleaching times and laser powers were tested on both types of car paints. Solid paint samples showed the best improvement with 10 % laser power while metallic samples worked the best with 50 % to 100 % laser power. Fluorescence was eliminated in 17 out of the 20 samples after photo bleaching was applied.

9.7.1.2 Art

Raman spectroscopy, as shown above, is helpful in pigment identification in car paints. This usefulness can also be extended to artistically relevant pigment identification. This information is used for art restoration purposes and to target forgeries of famous artwork. Specifically for forgeries, Raman spectroscopy is useful in identifying anachronistic pigments and identifying an artist's pigment profile. For a Raman spectroscopy review of general artistic supplies, the reader is referred to Centeno et al. [86], Slogget et al. [87] and Yu et al. [88] for reviews of art forensics before 2015.

Several studies aiming to authenticate specific pieces of art have been conducted. A Chagall painting was sampled with Raman spectroscopy by Chaplin et al. [89]. Fifteen spots on the painting, titled *Nude Woman Reclining*, were sampled with Raman spectroscopy at 632.8 nm. The analysis was able to identify seven pigments; of note were both phthalocyanine blue and phthalocyanine green. The painting had a suspected year of execution of 1910, but the two pigments in question were not developed until 1935. It was also shown that even after 1935, Chagall did not use phthalocyanine pigments on his paintings. The authors were able to conclude that the painting was not painted in 1910. A different painting called *Malatesta*, suspected to be painted in the Renaissance era, was investigated by Edwards et al. [90]. They used Raman spectroscopy to examine nine paint samples with two different wavelengths,

785 and 532 nm. The majority of pigments were found to correlate with pigments used during the Renaissance. However, chrome yellow, which was not developed until 1809, was also found in small areas of the painting, leading the authors to suppose it had been touched up in later years. Hibberts et al. [91] used a similar method to investigate a Renaissance painting in the *Noli Me Tangere* family. After investigation of 30 paint chips with Raman spectroscopy at 785 nm, all pigments were found to be consistent with the Renaissance time period (Figure 9.5).

De Faria et al. [92] used Raman spectroscopy at different wavelengths and other techniques to differentiate between two dyes called Indian yellow and tartrazine. Indian yellow is a historically relevant dye that has been used since the 1400's, but tartrazine is a synthetic yellow dye that was not produced until the 1800's. Fourier

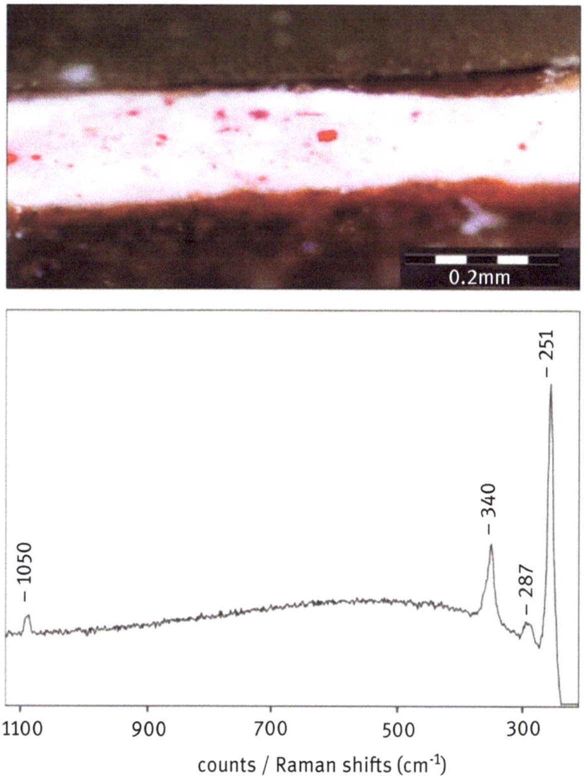

Figure 9.5: Mounted sample illustrating specimen stratigraphy, a flesh-pink sample; from the top, the degraded surface varnish, the white ground containing particles of red pigment, the subsurface orange resin and the underlying canvas can be clearly seen. Also shown here are the Raman spectra from the red and white areas, analyzing as cinnabar and lead white, respectively. *Reprinted from Hibberts, S., H. G. M. Edwards, et al. Raman spectroscopic analysis of a "noli me tangere" painting, Philosophical Transactions of the Royal Society A: Mathematical, Physical and Engineering Sciences 374, 2082 (2016) (ref. no. 91) Copyright 2016 with permission from The Royal Society.*

transform Raman spectroscopy was able to distinguish between the dyes only at 1064 nm because of fluorescent interference. A literature search revealed all previous identification of pigments as Indian yellow were incorrect and were actually tartrazine.

Pozzi et al. [93] investigated binary mixtures of red lake pigments using SERS. They found that certain red lake pigments are preferentially detected with SERS at 632.8 nm, even at lower concentrations. They also found little correlation between concentration and intensity of dyes in a mixture.

9.7.1.3 Other types of paint

Other types of paint, such as spray paint and house paint, while not as common, can be useful for forensic analysis. Lambert et al. [94] used several techniques combined with chemometrics to discriminate different house paints. Raman and IR spectra were taken of 34 red house paints and common component and specific weight analysis (CCSWA) was applied. With Raman data alone, the paints could only be separated into 7 groups but when IR and Raman spectroscopy were combined, 33 groups were able to be distinguished.

Spray paint ageing was investigated by Jost et al. [95] using both Raman and IR spectroscopy. Six different spray paints on glass slides were subjected to three different environments – an artificial climate chamber, the outdoors, and a dark room, which was used as a control. The Raman spectra did not change much over time because the Raman spectra is dominated by the signal from the pigments. The IR spectra, however, did show many changes over time due to oxidation.

Germinario et al. [96] used a variety of techniques and wavelengths to investigate spray paints. Raman spectroscopy was used specifically to identify pigments, with IR and pyrolysis GC added for binder identification. The Raman spectroscopic excitation wavelength utilized was varied according to color of paint, with 532 nm used for white, 638 nm used for red, yellow, brown, and black and 785 nm used for green and blue. When all the techniques were combined, the pigment and binders of all 45 paints were identified.

9.7.2 Fibers

A wide range of fibers and textiles are included in forensic fiber analysis, including natural fibers from both plant and animal sources, as well as semi-synthetic, and fully synthetic fibers. These fibers are then made into a wide range of products such as rope, automotive upholstery, rugs, and clothing [97]. Fiber analysis starts with the identification of the fiber type as either natural or manmade, using light or polarized light microscopy. A variety of techniques including FT-IR, Raman spectroscopy, TLC, microspectrophotometry, and HPLC are then used to identify the dye used [98]. Fiber evidence is at its most useful when there is a pool of knowns that are compared to the unknown in an effort to find a possible match. For a much more in-depth look into forensic fiber identification the reader is referred to *Forensic Examination of Fibres* [99] and a pre–2015 review by Meleiro et al. [100] on forensic analysis of textiles and fibers.

Bianchi et al. [101] studied the aging of fibers using Raman spectroscopy and statistical analysis. Cotton, polyester, and polyamide fibers underwent cycles of washes and UV light exposure to simulate ageing. New samples were washed only once while used and old samples were washed once, exposed to UV radiation, and then washed either an additional 6 times for used samples or 12 times for old. Both PCA and linear discriminant analysis (LDA) were used in an attempt to differentiate between the same type of fiber colored with different dyes, as well as differently aged samples of the same fiber. LDA models worked well for differentiating the same fibers with different dyes and the same fibers of different ages. Kavkler et al. [102] also investigated ageing, but focused specifically on bacterial induced ageing. Six different species of bacteria, isolated from old textiles, were used to inoculate cotton. After investigation with Raman spectroscopy at an excitation of 785 nm, certain species were found to cause structural changes to the cotton, including hydrolysis and depolymerisation.

Buzzini et al. [103] investigated the usefulness of various fiber analysis methods when compared to Raman spectroscopic analysis. A total of 180 blue, black, and red fibers of acrylic, cotton, and wool samples were used. Raman spectroscopic analysis was performed at 514.5, 632.8, 785, and 830 nm along with light microscopy, FT-IR, UV-vis microspectroscopy, and thin-layer chromatography. Discrimination power of all techniques and wavelengths individually and combined was analyzed. Raman spectroscopy by itself did not give satisfactory discrimination and for the best discrimination between fibers, multiple techniques were needed.

Was-Gubala et al. [104] investigated dye identification in low dye concentration mixtures using Raman and UV-vis microspectroscopy. Dye weight percentages ranged between 0.019 % and 3 % in both cotton and polyester samples. Excitation wavelengths of 785 nm or 514 nm were preferential over excitation at 633 nm. Binary and tertiary mixtures were especially complicated, due to domination of the spectra by the dye with the highest concentration. Only one dye could be seen in 61 % of tertiary dyed cotton fibers, 56 % of binary dyed cotton fibers, and 100 % of polyester fibers.

9.7.3 Hair

Hair is mostly made of protein and grows out of the skin of most mammals [105]. Head hair and pubic hair are the most forensically relevant types of hair, although hair can be found all over the body [106]. Forensic hair analysis uses different microscopy techniques at a variety of magnifications to observe visual characteristics of an unknown hair [107]. These characteristics are then compared to known samples. Pyrolysis GC, GC-MS, HPLC, trace element analysis, and DNA analysis can also be used on hair evidence; however, these techniques are usually used to investigate possible drug use by the donor or to identify different types of hair treatments [108]. There are two ways that hair can be transferred: directly and indirectly. Direct transfer is from a suspect to a victim, while indirect transfer is from a suspect to another object and then to a victim [109]. Because hair is so easily transferred, hair evidence can help provide associations, but cannot definitively identify an individual suspect [76]. More in-depth information concerning forensic hair investigation can be found in the text *Forensic Examination of Hair* [110]. A compendium of information regarding forensic hair investigation prior to 2015 can also be found in the work of Doty et al. [3].

The use of Raman spectroscopy is not common in forensic hair analysis, but because of its high specificity and easy sample preparation, it has great potential to be incorporated into the investigation. Hair is made primarily of keratin, with keratin associated proteins making up the majority of the interior of the hair. The majority of the physical properties of hair can be attributed to the high sulfur content and the large number of disulfide bonds present in keratin-associated proteins [111]. Fedorkova et al. [112] studied the formation of thiols in hair after weathering using Raman spectroscopy. Weathering was mimicked by exposing the hair to UV light. Gel-chromatography, UV-Vis spectrometry, SEM, light microscopy, and Raman spectroscopy were used to evaluate the cross sections of unpigmented hairs. UV radiation was found to damage the cuticle and increase the number of thiol groups found in the shaft of the hair. Raman spectra, in particular, indicated that there was a transformation of the type of disulfide bounds formed. The disulfide bonds changed from all gauche-gauche-gauche (ggg) bonds to a mixture of trans-gauche-gauche (tgg), and trans-gauche-trans (tgt) bonds.

Disulfide bond structure in hair was also examined by Kuzuhara et al. [113] Raman spectroscopy was used to examine how chemical hair waving, in conjunction with stress relaxation, effects keratin structure. In a conventional permanent waving treatment, disulfide bonds are reduced and then oxidized. New permanent waving techniques have emerged in which a hot or cold relaxation is added between the reduction and oxidation steps. Small cross sections of virgin white Chinese human hairs were subjected to different treatments and then interrogated with Raman spectroscopy at 514.5 nm. They found that the chemical used for reduction only reacts with the ggg and ggt conformations of said bonds and not with the tgt conformations. After the reduction, a high temperature relaxation increased the rate of sulfur bond breakage. A low temperature relaxation, on the other hand, increased the rate of sulfur bond formation. The high temperature relaxation between the reduction and oxidation damaged the hair significantly, as opposed to the low temperature relaxation.

Galvan et al. [114] evaluated different excitation wavelengths on their ability to distinguish between two different types of melanin using Raman spectroscopy. The two types of melanin, eumelanins and phemelanins, produce the observable differences seen in hair color between individuals of a species. Bird feathers, which are difficult to study non-invasively, were irradiated using wavelengths at 532, 780, and 1064 nm in an attempt to differentiate the two types of melanin. Only Raman spectroscopy excitation at 780 nm showed differences in the spectra of eumalamin and phemalanin. Excitation at 532 nm and 1064 nm were too high and too low in energy, respectively.

Wu et al. [115] used Raman spectroscopy to map high concentrations of squalene in hair. Squalene is a chemical compound in sebum, an oily substance that the skin produces naturally. Maps were made of microtomed hair cross sections using 785 nm Raman spectroscopy, and results showed squalene throughout the cross section in small discrete pockets.

9.8 Concluding remarks

Forensic science has become more and more popular in the public mind, and with this popularity has come a need to strengthen its scientific validity. Raman spectroscopy, with its distinct advantages including high specificity and nondestructive nature, has shown its usefulness for criminal investigations. Forensically relevant samples are very diverse and can encompass everything from paint chips to blood to gunshot residue. Raman spectroscopy is able to work with this infinite amount of sample types and can be as useful dealing with body fluid identification as it is with questioned documents. With both forensics and Raman spectroscopy on the rise, it will be interesting to see how their relationship will continue to unfold.

Funding: This project was supported by Award No. 2014-DN-BX-K016 and Award No. 2016-DN-BX–0166 awarded by the National Institute of Justice, Office of Justice Programs, U.S. Department of Justice. The opinions, findings, and conclusions or recommendations expressed in this publication are those of the authors and do not necessarily reflect those of the U.S. Department of Justice.

References

[1] Inman K, Rudin N. Chapter 3: overview- A unifying Paradigm of Forensic Science. In: principles and Practice of Criminalistics. Boca Raton FL: CRC Press LLC, 2001.

[2] Commitee on Identifying the Needs of the Forensic Sciences Commmunity. Strengthening Forensic Science in the United States: A Path Forward. National Academy of Science, 2009.

[3] Doty KC, Muro CK, Bueno J, Halámková L, Lednev IK. What can Raman spectroscopy do for criminalistics?. J Raman Spectrosc. 2016;47:39–50.

[4] Muro CK, Doty KC, Bueno J, Halámková L, Lednev IK. Vibrational spectroscopy: recent developments to revolutionize forensic science. Anal Chem. 2015;87:306–27.
[5] Doty KC, Lednev IK. Raman spectroscopy for forensic purposes: recent application for serology and gunshot residue. Trends Anal Chem. 2017. DOI: 10.1016/j.trac.2017.12.003
[6] Salahioglu F, Went MJ, Gibson SJ. Application of Raman spectroscopy for the differentiation of lipstick traces. Anal Methods. 2013;5:5392–401.
[7] Mohamad Asri MN, Mat Desa WNS, Ismail D. Raman spectroscopy of ballpoint-pen inks using chemometric techniques. Aust J Forensic Sci. 2017;49:175–85.
[8] Mistek E, Halámková L, Doty KC, Muro CK, Lednev IK. Race differentiation by Raman spectroscopy of a bloodstain for forensic purposes. Anal Chem. 2016;88:7453–56.
[9] Muro CK, Doty KC, De Souza Fernandes L, Lednev IK. Forensic body fluid identification and differentiation by Raman spectroscopy. Forensic Chem. 2016;1:31–8.
[10] Doty KC, McLaughlin G, Lednev IK, Raman A. “spectroscopic clock” for bloodstain age determination: the first week after deposition. Anal Bioanal Chem. 2016;408:3993–4001.
[11] Doty KC, Muro CK, Lednev IK. Predicting the time of the crime: bloodstain aging estimation for up to two years. Forensic Chem. 2017;5:1–7.
[12] Li R. Forensic serology. In: Kobilinsky LF, editor. Forensic chemistry handbook. 1st ed. Hoboken, N.J.: John Wiley & Sons, 2012:269–90.
[13] Rinke-Kneapler CN, Sigman ME. Applications of laser spectroscopy in forensic science. In: Baudelet M, editor. Laser spectroscopy for sensing: fundamentals, techniques and applications. Cambridge, UK: Woodhead Publishing, 2014:461–95.
[14] Virkler K, Lednev IK. Raman spectroscopy offers great potential for the nondestructive confirmatory identification of body fluids. Forensic Sci Int. 2008;181:e1–e5.
[15] Virkler K, Lednev IK. Raman spectroscopic signature of semen and its potential application to forensic body fluid identification. Forensic Sci Int. 2009;193:56–62.
[16] Virkler K, Lednev IK. Raman spectroscopic signature of blood and its potential application to forensic body fluid identification. Anal Bioanal Chem. 2010;396:525–34.
[17] Virkler K, Lednev IK. Forensic body fluid identification: the Raman spectroscopic signature of saliva. Analyst. 2010;135:512–17.
[18] Sikirzhytski V, Sikirzhytskaya A, Lednev IK. Multidimensional Raman spectroscopic signature of sweat and its potential application to forensic body fluid identification. Anal Chim Acta. 2012;718:78–83.
[19] Sikirzhytskaya A, Sikirzhytski V, Lednev IK. Raman spectroscopic signature of vaginal fluid and its potential application in forensic body fluid identification. Forensic Sci Int. 2012;216:44–8.
[20] Sikirzhytskaya A, Sikirzhytski V, Lednev IK. Raman spectroscopy coupled with advanced statistics for differentiating menstrual and peripheral blood. J Biophotonics. 2014;7:59–67.
[21] McLaughlin G, Doty KC, Lednev IK. Raman spectroscopy of blood for species identification. Anal Chem. 2014;86:11628–33.
[22] McLaughlin G, Sikirzhytski V, Lednev IK. Circumventing substrate interference in the Raman spectroscopic identification of blood stains. Forensic Sci Int. 2013;231:157–66.
[23] McLaughlin G, Lednev IK. A modified Raman multidimensional spectroscopic signature of blood to account for the effect of laser power. Forensic Sci Int. 2014;240:88–94.
[24] Zou Y, Xia P, Yang F, et al. Whole blood and semen identification using mid-infrared and Raman spectrum analysis for forensic applications. Anal Methods. 2016;8:3763–7.
[25] Feine I, Gafny R, Pinkas I. Combination of prostate-specific antigen detection and micro-Raman spectroscopy for confirmatory semen detection. Forensic Sci Int. 2017;270:241–7.
[26] Atkins CG, Buckley K, Blades MW, Turner RFB. Raman spectroscopy of blood and blood components. Appl Spectrosc. 2017;71:767–93.

[27] Bai P, Wang J, Yin H, Tian Y, Yao W, Gao J. Discrimination of human and nonhuman blood by Raman spectroscopy and partial least squares discriminant analysis. Anal Lett. 2017;50:379–88.
[28] Muro CK, Lednev IK. Race differentiation based on Raman spectroscopy of semen traces for forensic purposes. Anal Chem. 2017;89:4344–8.
[29] Sikirzhytskaya A, Sikirzhytski V, Lednev IK. Determining gender by Raman spectroscopy of a bloodstain. Anal Chem. 2017;89:1486–92.
[30] Muro CK, De Souza Fernandes L, Lednev IK. Sex determination based on Raman spectroscopy of saliva traces for forensic purposes. Anal Chem. 2016;88:12489–93.
[31] Muro CK, Lednev IK. Identification of individual red blood cells by Raman microspectroscopy for forensic purposes: in search of a limit of detection. Anal Bioanal Chem. 2017;409:287–93.
[32] United Nations Office on Drugs and Crime. World Drug Report 2017. 2017.
[33] Drug Enforcement Administration. Drugs of Abuse: A DEA Resource Guide, 2017 ed. U.S. Department of Justice, 2017.
[34] Assi S, Guirguis A, Halsey S, Fergus S, Stair JL. Analysis of 'legal high' substances and common adulterants using handheld spectroscopic techniques. Anal Methods. 2015;7:736–46.
[35] Penido CAFO, Pacheco MTT, Zângaro RA, Silveira L. Identification of Different Forms of Cocaine and Substances Used in Adulteration Using Near-infrared Raman Spectroscopy and Infrared Absorption Spectroscopy. J Forensic Sci. 2015;60:171–8.
[36] Bono JP. Criminalistcs: introduction to Controlled Substances. In: editor, Karch SB. Drug Abuse Handbook. 2nd ed. Boca Raton, FL: Taylor & Francis Group, LLC, 2006.
[37] Rebiere H, Ghyselinck C, Lempereur L, Brenier C. Investigation of the composition of anabolic tablets using near infrared spectroscopy and Raman chemical imaging. Drug Test Anal. 2016;8:370–7.
[38] Jones LE, Stewart A, Peters KL, et al. Infrared and Raman scattering of seized novel psychoactive substances: a large scale study of >200 samples. Analyst. 2016;141:902–9.
[39] Elie L, Elie M, Cave G, Vetter M, Croxton R, Baron M. Microcrystalline testing used in combination with Raman micro-spectroscopy for absolute identification of novel psychoactive substances. J Raman Spectrosc. 2016;47:1343–50.
[40] Mackey TK, Liang BA. The global counterfeit drug trade: patient safety and public health risks. J Pharm Sci. 2011;100:4571–9.
[41] Rojek C. Counterfeit Commerce: relations of Production, Distribution and Exchange. Cult Sociol. 2017;11(1):28–43.
[42] Mukhopadhyay R. The hunt for counterfeit medicine. Anal Chem. 2007;79:2622–7.
[43] Aboul-Enein Y, Bunaciu AA, Fleschin S. Spectroscopic Analytical Methods for Detection of Counterfeit Pharmaceutical Preparations–A Mini-Review. Gazi Univ J Sci. 2013;26:407–17.
[44] Loethen YL, Kauffman JF, Buhse LF, Rodriguez JD. Rapid screening of anti-infective drug products for counterfeits using Raman spectral library-based correlation methods. Analyst. 2015;140:7225–33.
[45] Kwok K, Taylor LS. Analysis of the packaging enclosing a counterfeit pharmaceutical tablet using Raman microscopy and two-dimensional correlation spectroscopy. Vib Spectrosc. 2012;61:176–82.
[46] Molina DK. Chapter 2: methodology. In: Handbook of Forensic Toxicology for Medical Examiners. Boca Raton FL: Taylor & Francis Group LLC, 2009.
[47] Kronstrand R, Jones AW. Drugs of Abuse/Analysis. In: Siegel J, Knupfer G, Saukko P, editors. Encyclopedia of Forensic Sciences. 2000:598–610.
[48] Bumbrah GS, Sharma RM. Raman spectroscopy – basic principle, instrumentation and selected applications for the characterization of drugs of abuse. Egypt J Forensic Sci. 2016;6:209–15.

[49] D'Elia V, Montalvo G, Ruiz CG. Analysis of street cocaine samples in nasal fluid by Raman spectroscopy. Talanta. 2016;154:367–73.
[50] Bell S. Explosives. In: Forensic Chemistry, 2ndед. Pearson Education Inc., 2013.
[51] Zapata F, López-López M, García-Ruiz C. Detection and identification of explosives by surface enhanced Raman scattering. Appl Spectrscopy Rev. 2016;51:227–62.
[52] Zapata F, Fernández De La Ossa Á, Gilchrist E, Barron L, García-Ruiz C. Progressing the analysis of Improvised Explosive Devices: comparative study for trace detection of explosive residues in handprints by Raman spectroscopy and liquid chromatography. Talanta. 2016;161:219–27.
[53] Matyáš R, Lyčka A, Jirásko R, et al. Analytical Characterization of Erythritol Tetranitrate, an Improvised Explosive. J Forensic Sci. 2016;61:759–64.
[54] Zapata F, García-Ruiz C. Determination of Nanogram Microparticles from Explosives after Real Open-Air Explosions by Confocal Raman Microscopy. Anal Chem. 2016;88:6726–33.
[55] Bueno J, Sikirzhytski V, Lednev IK. Raman Spectroscopic Analysis of Gunshot Residue Offering Great Potential for Caliber Differentiation. Anal Chem. 2012;84:4334–9.
[56] Bueno J, Lednev IK. Raman microspectroscopic chemical mapping and chemometric classification for the identification of gunshot residue on adhesive tape. Anal Bioanal Chem. 2014;406:4595–9.
[57] Bueno J, Lednev IK. Advanced statistical analysis and discrimination of gunshot residue implementing combined Raman and FT-IR data. Anal Methods. 2013;5:6292–6.
[58] López-López M, Fernández De La Ossa MÁ, García-Ruiz C. Fast Analysis of Complete Macroscopic Gunshot Residues on Substrates Using Raman Imaging. Appl Spectrosc. 2015;69:889–93.
[59] Giles A. Forensic examination of documents. In: White PC, editor. Crime scene to court: the essentials of forensic science, Fourth. editor Cambridge, UK: The Royal Society of Chemistry, 2016:229–59.
[60] Vos M, Strach S, Westwood S. Document analysis/handwriting. In: Siegel JA, Saukko PJ, editors. Encyclopedia of forensic sciences. Academic Press, 2000:584–90.
[61] Purdy DC. Document analysis/document dating. In: Siegel JA, Saukko PJ, editors. Encyclopedia of forensic sciences. Academic Press, 2000:570–80.
[62] Aginsky V. Document analysis/analytical methods. In: Siegel JA, Saukko PJ, editors. Encyclopedia of forensic sciences. Academic Press, 2000:566–70.
[63] Calcerrada M, García-Ruiz C. Analysis of questioned documents: a review. Anal Chim Acta. 2015;853:143–66.
[64] Buzzini P, Suzuki E. Forensic applications of Raman spectroscopy for the in situ analyses of pigments and dyes in ink and paint evidence. J Raman Spectrosc. 2016;47:16–27.
[65] Nunkoo MI, Saib-Sunassy MB, Li Kam Wah H, Jhaumeer Laulloo S. Forensic analysis of black, blue, red, and green ballpoint pen inks. In: Ramasami P, Bhowon MG, Laulloo SJ, Wah HLK, editors. Crystallizing ideas–the role of chemistry. Springer; 2016. p. 323–39.
[66] Lee LC, Samad MIA, Ismail MAM. Nondestructive classification and identification of ballpoint pen inks by Raman spectroscopy for forensic document examinations. J Anal Chem. 2016;71:723–9.
[67] Mohamad Asri NM, Mat Desa WNS, Ismail D. Source determination of red gel pen inks using Raman spectroscopy and attenuated total reflectance Fourier transform infrared spectroscopy combined with pearson's product moment correlation coefficients and principal component analysis. J Forensic Sci. 2018;63(1):285–91.
[68] Borba F, Jawhari T, Honorato RS, De Juan A. Confocal Raman imaging and chemometrics applied to solve forensic document examination involving crossed lines and obliteration cases by a depth profiling study. Analyst. 2017;142:1106–18.

[69] Huynh V, Williams KC, Golden TD, Verbeck GF. Investigation of falsified documents via direct analyte-probed nanoextraction coupled to nanospray mass spectrometry, fluorescence microscopy, and Raman spectroscopy. Analyst. 2015;140:6553–62.
[70] Johnson CE, Martin P, Roberts KA, et al. The capability of Raman microspectroscopy to differentiate printing inks. J Forensic Sci. 2018;63(1):66–79. DOI:.
[71] Chen R, Lv J, Feng J, Liu Y, Zhang W. Discrimination of seal inks used for seals by confocal Raman microscopy. Pigm Resin Technol. 2014;43:389–93.
[72] Alves APP, De Oliveira LPZ, Castro AAN, et al. The structure of different cellulosic fibres characterized by Raman spectroscopy. Vib Spectrosc. 2016;86:324–30.
[73] Zięba-Palus J, Wesełucha-Birczyńska A, Trzcińska B, Kowalski R, Moskal P. Analysis of degraded papers by infrared and Raman spectroscopy for forensic purposes. J Mol Struct. 2017;1140:154–62.
[74] Bell S. Chemical Analysis of Materials. In: Forensic Chemistry, 2nd ed. Pearson Education, 2013.
[75] Bentley J. Composition, Manufacture and Use of Paint. In: Caddy B, editor. Forensic Examination of Glass and Paint. Analysis and Interpretation CRC Press, 2001.
[76] Maxwell VM. Forensic Examination of Trace Evidence. In: Forensic Science. Wiley-VCH Verlag GmbH & Co. KGaA, 2016:337–71.
[77] Henson ML, Jergovich TA. Scanning electron microscopy and energy dispersive X-ray spectrometry (SEM/EDS) for the forensic examination of paints and coatings. In: Caddy B, editor. Forensic Examination of Glass and Paint. Analysis and Interpretation. CRC Press, 2001.
[78] Forensic Examination of Glass and Paint. Analysis and Interpretation CRC Press, 2001.
[79] Ferreira KB, Oliveira AGG, Gomes JA. Raman spectroscopy of automotive paints: forensic analysis of variability and spectral quality. Spectrosc Lett. 2017;50:102–10.
[80] Chen R, Lv J, Feng J. Characterization of Paint by Fourier-Transform Infrared Spectroscopy, Raman Microscopy, and Scanning Electron Microscopy-Energy Dispersive X-ray Spectroscopy. Anal Lett. 2015;48:1502–10.
[81] Lv JG, Liu S, Feng JM, Liu Y, Zhou SD, Chen R. Effective identification of paints pigments in hit-and-run cases with confocal Raman microscope. Pigment Resin Technol. 2016;45:294–300.
[82] Lv J, Zhang W, Liu S, et al. Analysis of 52 sutomotive couting samples for forensic purposes with Fourier transform infrared spectroscopy (FTIR) and Raman microscopy. Environ Forensics. 2016;17:59–67.
[83] Martyna A, Michalska A, Zadora G. Interpretation of FTIR spectra of polymers and Raman spectra of car paints by means of likelihood ratio approach supported by wavelet transform for reducing data dimensionality. Anal Bioanal Chem. 2015;407:3357–76.
[84] Michalska A, Martyna A, Zięba-Palus J, Zadora G. Application of a likelihood ratio approach in solving a comparison problem of Raman spectra recorded for blue automotive paints. J Raman Spectrosc. 2015;46:772–83.
[85] Zięba-Palus J, Michalska A. Photobleaching as a useful technique in reducing of fluorescence in Raman spectra of blue automobile paint samples. Vib Spectrosc. 2014;74:6–12.
[86] Centeno SA. Identification of artistic materials in paintings and drawings by Raman spectroscopy: some challenges and future outlooks. J Raman Spectrosc. 2016;47:9–15.
[87] Soggett R. Art crime: fraud and forensics. Aust J Forensic Sci. 2015;47:253–59.
[88] Yu J, Butler IS. Recent applications of infrared and Raman spectroscopy in art forensics: A brief overview. Appl Spectrosc Rev. 2015;50:152–7.
[89] Chaplin TD, Clark RJH. Identification by Raman microscopy of anachronistic pigments on a purported Chagall nude: conservation consequences. Appl Phys A Mater Sci Process. 2016;122:1–5.
[90] Edwards HGM, Vandenabeele P, Benoy TJ. Raman spectroscopic study of “the Malatesta”: A Renaissance painting?. Spectrosc Acta A Mol Biomol Spectrosc. 2015;137:45–49.

[91] Hibberts S, Edwards HGM, Abdel-Ghani M, Vandenabeele P. Raman spectroscopic analysis of a 'noli me tangere' painting. Philosl Trans Royal Soc A Mathemat Phys Eng Sci. 2016;374:10.
[92] De Faria DLA, Edwards HGM, Careaga V, Walt N, Maier MS. A definitive analytical spectroscopic study of Indian yellow, an ancient pigment used for dating purposes. Forensic Sci Int. 2017;271:1–7.
[93] Pozzi F, Zaleski S, Casadio F, Van Duyne RP. SERS Discrimination of Closely Related Molecules: A Systematic Study of Natural Red Dyes in Binary Mixtures. J Phys Chem C. 2016;120:21017–26.
[94] Lambert D, Muehlethaler C, Esseiva P, Massonnet G. Combining spectroscopic data in the forensic analysis of paint: application of a multiblock technique as chemometric tool. Forensic Sci Int. 2016;263:39–47.
[95] Jost C, Muehlethaler C, Massonnet G. Forensic aspects of the weathering and ageing of spray paints. Forensic Sci Int. 2016;258:32–40.
[96] Germinario G, Van Der Werf ID, Sabbatini L. Chemical characterisation of spray paints by a multi-analytical (Py/GC–MS, FTIR, μ-Raman) approach. Microchem J. 2016;124:929–39.
[97] David SK, Pailthorpe MT. Classification of Textile Fibres: production, Structure, and Properties. In: Robertson J, Grieve M, editor. Forensic Examination of Fibres, 2 ed. CRC Press, 1999.
[98] Palmer R. Fibers: identification and Comparison. In: Siegel JA, Saukku PJ, editors. Encyclopedia of Forensic Sciences. Academic Press, 2000:815–23.
[99] Forensic Examination of Fibres, 2 ed. CRC Press, 1999.
[100] Meleiro PP, García-Ruiz C. Spectroscopic techniques for the forensic analysis of textile fibers. Appl Spectrosc Rev. 2016;51:278–301.
[101] Bianchi F, Riboni N, Trolla V, et al. Differentiation of aged fibers by Raman spectroscopy and multivariate data analysis. Talanta. 2016;154:467–73.
[102] Kavkler K, Gunde Cimerman N, Zalar P, Demšar A. Deterioration of contemporary and artificially aged cotton by selected fungal species. Polym Degrad Stab. 2015;113:1–9.
[103] Buzzini P, Massonnet G. The analysis of colored acrylic, cotton, and wool textile fibers using micro-raman spectroscopy. Part 2: comparison with the traditional methods of fiber examination. J Forensic Sci. 2015;60:712–20.
[104] Was-Gubala J, Starczak R. Nondestructive identification of dye mixtures in polyester and cotton fibers using Raman spectroscopy and ultraviolet-visible (UV-Vis) microspectrophotometry. Appl Spectrosc. 2015;69:296–303.
[105] Gaudette BD. Hair Overview. In: Siegel J, Saukko P, editors. Encyclopedia of Forensic Sciences Academic Press. 2000:999–1041.
[106] Bisbing RE. Hair: comparison Microscopic. In: Siegel JA, Saukko PJ, editor. Encycopedia of Forensic Sciences, 1st ed. Academic Press, 2000.
[107] Robertson J. Forensic and Microscopic Examination of Human Hair. In: Robertson J, editor. Forensic Examination of Hair. CRC Press, 1999:79–154.
[108] Gaudette BD. Hair: comparision Other. In: Siegel JA, Suakko P, editors. Encycopedia of Forensic Sciences. Academic Press. 2000:1016–18.
[109] Gaudette BD. Evidential Value of Hair Examination. In: Robertson J, editor. Forensic Examination of Hair. CRC Press, 1999:229–42.
[110] Forensic Examination of Hair. CRC press, 1999.
[111] Harding H, Rogers G. Physiology and Growth of Human Hair. In: Robertson J, editor. Forensic Examination of Hair. CRC Press, 1999:1–78.
[112] Fedorkova MV, Brandt NN, Chikishev AY, et al. Photoinduced formation of thiols in human hair. J Photochem Photobiol B: Biol. 2016;164:43–8.

[113] Kuzuhara A. Internal structural changes in keratin fibres resulting from combined hair waving and stress relaxation treatments: a Raman spectroscopic investigation. Int J Cosmet Sci. 2016;38:201–09.
[114] Galván I, Jorge A. Dispersive Raman spectroscopy allows the identification and quantification of melanin types. Ecol Evol. 2015;5:1425–31.
[115] Wu Y, Chen G, Ji C, et al. Gas chromatography-mass spectrometry and Raman imaging measurement of squalene content and distribution in human hair. Anal Bioanal Chem. 2016;408:2357–62.

Index

https://doi.org/10.1515/9783110515312-010

Bei Fragen zur Produktsicherheit wenden Sie sich bitte an:
If you have any questions regarding product safety,
please contact:

Walter de Gruyter GmbH
Genthiner Straße 13
10785 Berlin
productsafety@degruyterbrill.com